CNC Robotics

Build Your Own
Workshop Bot

Geoff Williams

McGraw-Hill

New York   Chicago   San Francisco   Lisbon
London   Madrid   Mexico City   Milan
New Delhi   San Juan   Seoul
Singapore   Sydney   Toronto

The McGraw·Hill Companies

Cataloging-in-Publication Data is on file with the Library of Congress

4 5 6 7 8 9 0   DOC/DOC   0 9 8 7 6 5

ISBN 0-07-141828-8

The sponsoring editor for this book was Judy Bass and the production supervisor was Pamela Pelton. It was set in Tiepolo Book by Patricia Wallenburg.

Printed and bound by RR Donnelly.

McGraw-Hill books are available at special quantity discounts to use as premiums and sales promotions, or for use in corporate training programs. For more information, please write to the Director of Special Sales, McGraw-Hill Professional, Two Penn Plaza, New York, NY 10121-2298. Or contact your local bookstore.

 This book is printed on recycled, acid-free paper containing a minimum of 50 percent recycled, de-inked fiber.

For Margaret, whose help and patience made this book possible.

Contents

Acknowledgments

I must thank my brother Karl who inspired me to write this book and my editor Judy Bass whose faith and assistance made the book a reality. I'd also like to thank Patricia Wallenburg who assembled my words and images into book form. Judy and Patricia have made this book project an extremely enjoyable experience. Finally my thanks go out to all the people who have freely shared with me their knowledge and assistance while I was researching and building my CNC machine.

I must thank the following companies for allowing me to include some of their copyrighted material in this book.

The NuArc Company, Inc. doesn't promote, endorse, or warranty any modified products. NuArc let me reproduce some of the images from the repair manual of the copy camera I disassembled but they don't endorse the use of their products for anything other than their originally intended function. You can contact NuArc at M&R Sales and Service, Inc. 1 N. 372 Main Street, Glen Ellyn, IL 60137, USA or on the Web at http://www.nuarc.com.

Kellyware has allowed me to use screen captures of the program KCam 4. Kellyware can be contacted at PO Box 563, Spirit Lake, Iowa 52360, USA or on the web at http://www.kellyware.com.

The product data sheets included with Chapter 2 of this book have been reprinted with the permission of STMicroelectronics. The documents reproduced in this book and many more useful application notes can be found at the STMicroelectronics Web site located at http://www.st.com.

Science Specialists, Inc. has given me permission to include screen captures of the software ACME Profiler, Coyote Version 6.0. Science Specialists, Inc. can be reached at 1800 Sheckler Rd., Columbia City, IN 4675, USA or on the Web at http://www2.fwi.com/ ~ kimble/ scispec/scispec.htm.

1
Design

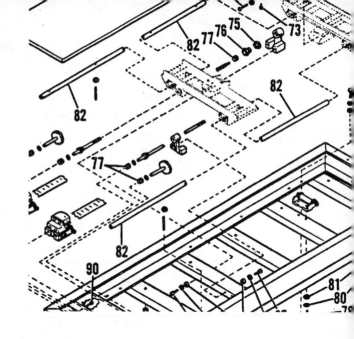

Why Build My Own

I first thought about adding a CNC router to my tool collection after finishing a kitchen cabinet renovation in my home. I refaced the cabinets and built 26 new doors, during which I discovered that door building can become monotonous at best. As always happens when you tell or show your friends and family what you have done, someone will have a similar project and enlist your help. That someone was my friend Geoff S. He wanted to do the same thing to his kitchen—reface and install new cabinet doors. I agreed to help him and he decided on a style of door that can be made from one piece of material cut to size and routed to create the look he wanted. Of course the prospect of building a whole lot of doors and making templates to facilitate the routing wasn't too thrilling. That's when I thought a small CNC machine would come in handy. All the repetitive routing could be assigned to the CNC machine and the doors would more closely resemble each other once human error had been removed from the equation. Now the project didn't seem too bad at all!

I started to look for an affordable machine to do the job, After searching the Internet, I was shocked to find how much the asking price is for a CNC machine. I did find a couple that were under

$6000 U.S., but I can never convince myself to buy a tool worth so much. Even if the cost of the machine seems reasonable you still have shipping and duties to pay, and in my case the exchange rate between U.S. and Canadian dollars. All things considered, it was going to cost me in excess of $10,000 Canadian to get a CNC machine in my shed. I can't afford that kind of price tag! I searched for plans or a book that described what I wanted to build. I did find some plans on the Internet but either the machine was too small and inaccurate or the plans were expensive and required the use of expensive components. I couldn't find any books in print about a similar project. I won't buy plans that I can't get a good look at first, so the Internet plans were out of the question. I prefer books because I can hold them and flip through the pages before I hand over the cash. Books also cost less.

I decided to build my own machine using some off-the-shelf lineal motion components and some components that I salvaged or modified to suit the project. I thought the most logical thing would be to document my progress and share the information through a book. To summarize, I decided to build my own machine because I love a challenge and I learn more when I have a practical project; also, I can keep the cost low. It's that simple.

Gantry Style

In my opinion, a gantry styled CNC machine is simplest to implement. A few years ago, I built a band saw mill frame and gantry, so the design of a more accurate system didn't seem too tough a project. I also like the idea of moving the tool over the material rather than the material under the tool. A machine built to move material would not have as large a working area for a given footprint. Considering my work shed is only 22 X 12 feet, a gantry machine is most suitable.

Motors

The first purchase to make was the stepper motor. My local Princess Auto has a great surplus department, so I headed there first. Sure enough, they had some step-syn motors (seen in **Figures** 1.1 and 1.2). They are Nema frame size 34, draw 1.4 amps per channel, and have a rating of 4.6 volts.

Figure 1.1

Step-Syn stepper motor side view.

Figure 1.2

Step-Syn stepper motor top view.

These motors were used in an IBM product—probably a printer. They are unipolar, but if you run them as bipolar they produce more torque (see **Figure** 1.3).

I also discovered that these stepper motors work better if they are given 12 volts instead of the 4.6-volt rating on the motor body. The strength of a stepper motor is rated in ounce inches of holding torque. The step-syn information I found indicated that these motors are anywhere from 90 to 220 ounce inches. I sometimes work backwards, and buying the motors first is certainly just that! Normally, you would calculate what strength of motor you need to run the machine and then purchase a suitable motor. Here's how to calculate the strength of motor you need to run this machine. If you can't find any surplus motors, investigate a company called Pacific Scientific—they have a variety of stepper motors and also make available software for download, which you can use to determine the size of motor you need. Speaking with them, I was impressed with how well I was treated, considering I would only need three of their motors.

Remember that although brand new motors are expensive, you know they will work and you can match the strength to the machine. New motors could also speed up the machine considerably. When I say "speed up," it is important to note that I am referring to *travel* speeds, not cutting speeds. Cutting speeds for most materials will be slow with this style of machine, regardless of which motor you choose; you can't run a router through wood at 200 inches per minute and expect the cut to look good. Cutting speeds of 10 to 30 inches per minute define the range we can expect from this machine with these motors. A faster machine can be expected to travel quickly when not cutting and slow down when cutting through material.

Lineal Motion

There are a variety of off-the-shelf lineal motion products, but most of the systems were too expensive for this project. Thinking that it would be useful some day, I acquired a NuArc copy camera a few years ago (see **Figure** 1.4), so I took it apart and found it

STEP-SYN 103-820-0240 4.5V 1.4 AMP 2DEG/STEP
WIRING DIAGRAM

Figure 1.3

Wiring diagram of a
Step-Syn stepper motor.

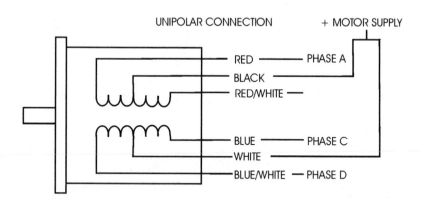

UNIPOLAR CONNECTION   + MOTOR SUPPLY

RED ——————— PHASE A
BLACK ———————
RED/WHITE ——
BLUE ——————— PHASE C
WHITE ———————
BLUE/WHITE — PHASE D

BIPOLAR CONNECTIONS

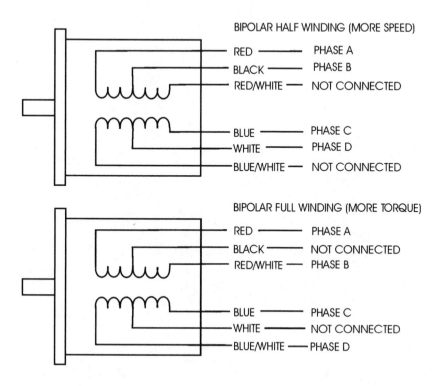

BIPOLAR HALF WINDING (MORE SPEED)

RED ——————— PHASE A
BLACK ——————— PHASE B
RED/WHITE — NOT CONNECTED
BLUE ——————— PHASE C
WHITE ——————— PHASE D
BLUE/WHITE — NOT CONNECTED

BIPOLAR FULL WINDING (MORE TORQUE)

RED ——————— PHASE A
BLACK ——————— NOT CONNECTED
RED/WHITE — PHASE B
BLUE ——————— PHASE C
WHITE ——————— NOT CONNECTED
BLUE/WHITE — PHASE D

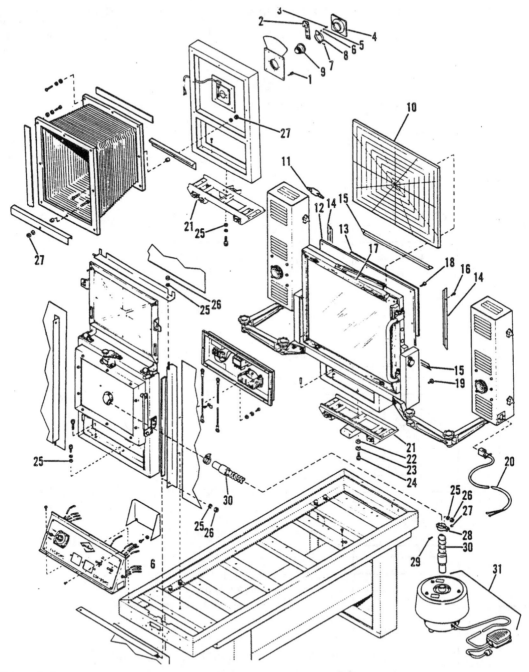

Figure 1.4

Exploded illustration of the NuArc Model SST 1418 supersonic horizontal camera. Part 21 is the carriage that travels on the guide rails using lineal bearings.

uses lineal bearings running on guide rails to move the copy board as well as the bellows.

The slide employed to move the bellows and copy board would work well as the y-axis for my CNC machine. And it was complete with bearings and holders built into the slide, as pictured in **Figures** 1.5 and 1.6.

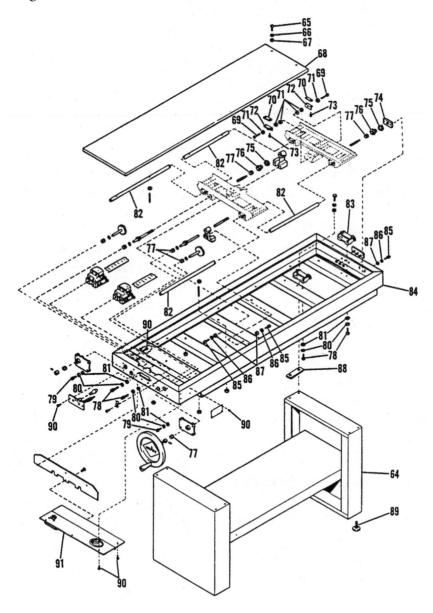

Figure 1.5

NuArc Camera, part number 82 is the guide rail.

Figure 1.6

Closeup of the slide showing the bearings numbered 40 and the location of lead screw nut numbered 33.

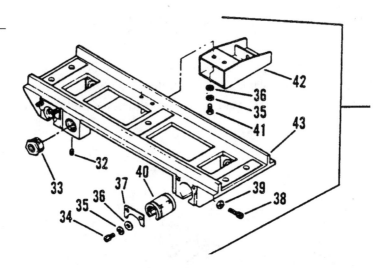

Of the eight bearings, I found that only four were still in satisfactory condition for use. I decided that the x-axis could be built in the same manner, employing open lineal bearings running on a rail that had been drilled and tapped to allow the use of support bolts. I also chose to make my own bearing holders for the x-axis because the cost of prebuilt products was more than I could justify. Rail support material is available as well, but the cost of this product made me believe it wasn't required and that the bolts would give the rail enough support. I noticed that the copy camera didn't have any extra support under the rails. If you wanted extra support in a project like this, it could be fashioned from two pieces of angle iron with a spacer, but the surface it would be mounted on—considering the rail would be in contact with the support—would have to be perfectly flat. As I had no intention of using perfect steel to build this machine, having the bolts provide the support meant that they could be adjusted to bring the rail to a flat plane.

Similar bearings and rails could be used for the z-axis, but I decided instead to use a swiveling TV tray assembly bought at the Home Depot. The glides are rated at 100 lbs. to hold a television horizontally. My project would use the glides vertically, so they would be plenty strong, with large ball bearings and enough travel for the z-axis. The NuArc camera used 3/4-inch bearings and support rails, so I decided to use the same bearing and rail size on the x-axis.

Motor Drivers

At this point in the project I had already purchased motors, so I looked at simple driver solutions. The best solution was found in the form of the 1297 and 1298 integrated circuits manufactured by ST Microelectronics. Their Web site has all the information needed to build a bipolar stepper motor driver using these two integrated circuits, which are often referred to as "chips." A driver built from these chips can easily provide the voltage and amperage needed by the step syn motors.

Acme Screw

The question of how to move the gantry and axes slides was also resolved by cost. I had originally considered using ball screws, but after comparing the cost of the *ball screw* with that of an *acme screw*, it didn't make sense to spend three times as much on ball screws. The advantages to using ball screws are that a smaller motor can be used to move a given load, and with a preloaded nut, there is very little backlash in the system. As mentioned earlier in this chapter, this is a machine that will not speed through its assigned jobs so we can compensate for backlash in the software. This means the project can be built using less expensive acme screws.

I also had to decide how many turns per inch to put on the acme screw. My experiments with *ready rod* proved that too many turns made for annoyingly slow movement and too few turns reduces the quality of resolution that allows the machine to make small, precise movements. I settled on a 1/2-inch acme screw with eight turns per inch, and a 6-foot length with a nut at a cost of $135 Canadian.

Deciding on the Dimensions of the Machine

Earlier in the chapter I explained that I made the decision of machine footprint size based on the area in my workshop.

Because I only have a space 12 × 22 feet and tools and materials currently occupy most of that space, my machine would only be about 7 feet long and 4 feet wide. The next step in this project was to generate concept drawings, since I was going to use some of the components from the NuArc horizontal camera. The dimensions of the frame ended up being longer than the support rails in order to accommodate the bearing holders and the motor mount with a little room to spare. The width of the frame is a few inches shorter than the balance of a 6-foot acme lead screw, after the length needed for the z-axis has been cut from it. The following illustrations are the concept drawings I created to guide the construction of the machine. **Figure 1.7** is a drawing of the machine from the side.

Figure 1.7

View from side of proposed machine.

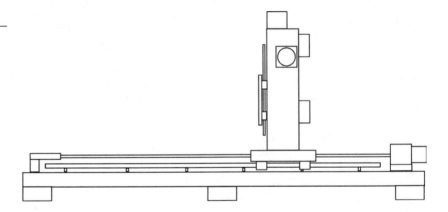

The next illustration, **Figure 1.8**, is the width of the machine, viewed from the front.

Software

After figuring out the approximate shape and dimensions and deciding on stepper motors and drivers, the next question was which software to use to control the machine once finished. I looked at a variety of software solutions and, since my level of experience with CNC machinery was nonexistent, I wanted a program that was easy to use. To communicate to the stepper motors

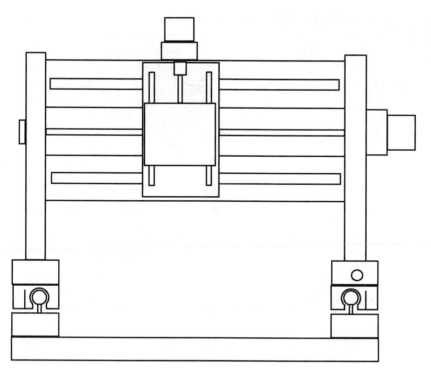

Figure 1.8

Front view of machine along its width.

how to move so the tool being used will follow the desired path, a program is written in *G-code* and *M-code*. The G- and M-codes are used to tell the machine where to go in the xyx-axes areas of travel and what to do when it gets there. Very simple programs describing things like boxes or circles are not very complicated to write yourself. I wanted software that would allow me to create my own designs in a drawing program like CorelDraw and then import the drawing and automatically create the necessary G- and M-code file. I looked at some freeware but was disappointed by the level of difficulty to implement the software and get it doing what I wanted. Further research revealed software called KCam that would do exactly what was necessary. See the screen capture from KCam in **Figure 1.9**.

KCam isn't freeware, but it certainly isn't expensive either at $100 U.S. per copy. The fact that KCam is not expensive shouldn't lead you to believe it is ineffective software. It is extremely easy to use and allows you to customize the use of the printer port. KCam will

Figure 1.9

What KCam looks like.

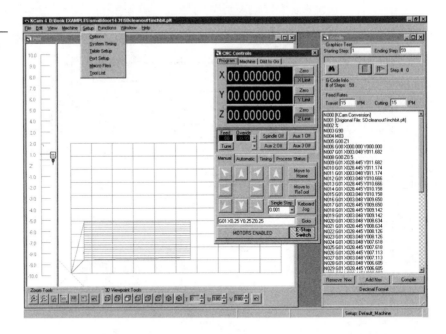

also import *HPGL files* created in CorelDraw 9 or *DXF files* created in CAD software like AutoCad or in CorelDraw 9.

In this chapter I shared design choices for my CNC machine and the reasons behind them, as well as the choices of stepper motors, drivers, and software. At this point in the project, we need to start thinking about the electronics, the topic of the next chapter.

2
Electronics

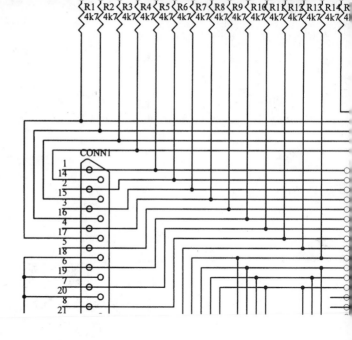

Stepper Motor Driver and Computer Interface Boards

This chapter deals with the design and construction of the electronics required for controlling stepper motors using a computer and KCam software. The requirements for the drivers are based on the step-syn motor purchased in the previous chapter from the surplus department of Princess Auto (see **Figure 2.2**).

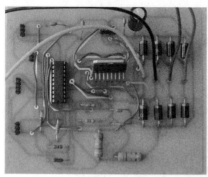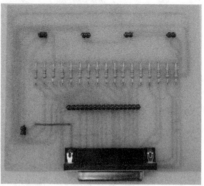

Figure 2.1

The finished boards.

In addition to these surplus motors, I also bought three new motors from Pacific Scientific, also mentioned in the previous chapter (see **Figure 2.3**).

Figure 2.2

Step-syn stepper motor.

Figure 2.3

Pacific Scientific
stepper motor.

All the motors draw 1.4 amps per channel, but the step-syn motors are rated at 4.5 volts and the Pacific Scientific motor can be given a maximum of 170 volts. Although I'm not using Pacific Scientific motors in this project, if you are unable to find suitable surplus steppers you may wish to purchase new ones. **WARNING:** The *stepper motor driver boards* described in this chapter can drive both of these motors, but you can *only use up to 36 volts for motor power*—too many volts will fry the L298 integrated circuit. A company called ST Microelectronics has graciously given me permission to reproduce the data sheets and application notes in their entirety in this chapter (see pages 22 through 75). These documents are the basis of the driver board design. Read them thoroughly to gain a greater understanding of the strengths and limitations of the L297/L298 integrated circuits (referred to as ICs) in this application.

Stepper Motor Driver Circuit

The *stepper motor driver boards* are the muscles of the CNC machine. They receive signals from the computer that indicate which direction that axis will travel and how far it will move. They are the muscles because as they receive direction and step signals from

Figure 2.4

L297 stepper motor controller IC.

Figure 2.5

L298 H Bridge IC.

the computer, they translate the information into higher voltage and amperage signals to send to the stepper motors. The power sent to the stepper motor coils is distributed to the coils in a sequence that will move the shaft in the desired direction as many steps as are needed to traverse the distance required on that axis. For this project we need three driver boards, one for each axis of travel. The boards are designed using a set of integrated circuits manufactured by STMicroelectrionics—the L297 and L298. The nice thing about using these two chips is that the board design is quite simple, only requiring a minimal number of components. A second benefit is that when combined, these two chips create a very powerful driver board capable of handling up to 36 volts and 2 amperes per channel. A lot of bipolar and uniploar stepper motors currently manufactured or available as surplus, which are strong enough to be used for this machine, are well within the tolerances of these chips. The steppers that I decided to use are Sanyo Denki step-syn and are rated at 4.5 volts and 1.4 amps per channel with a resolution of 2 degrees per step. The power ratings are well within the tolerances of the driver board. You can refer to the schematic (**Figure 2.6**) to determine the components required for this board.

Table 2-1	Part	Quantity	Description
Driver Board Components	U1	1	L298 Dual full-bridge driver
	U2	1	L297 Stepper motor controller
	D1–D8	8	FR304 Fast recovery diode
	C1	1	3.3 NF capacitor
	C2, C3	2	0.1 UF capacitor
	C4	1	470 UF capacitor
	C5, C6	2	1NF capacitor
	R1, R2	2	0.5 ohm power resistor
	R3	1	1K ohm resistor
	R4	1	22K ohm resistor

(continued on next page)

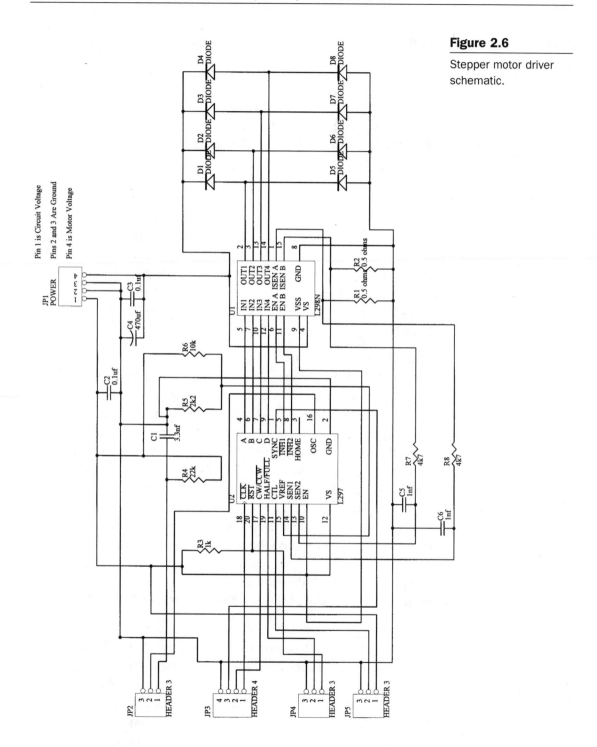

Figure 2.6

Stepper motor driver schematic.

Part	Quantity	Description
Table 2-1 Driver Board Components (continued)		
R5	1	2K2 ohm resistor
R6	1	10K ohm resistor
R7, R8	2	4.7K ohm resistor
JP1–JP6	3-3pin, 2-4pin	Cut to size from header material.
Heat Sink	1	You must install a heat sink on the L298.

Now for a brief explanation of the circuit; note that I am including in this chapter the data sheets for the L298 and L297 ICs as well as the application notes, so you will be able to refer to the source material at will (see pages 22 through 75). The information contained in these documents is essentially all you need to create the circuit you are about to build.

This circuit works by receiving signals from the computer's parallel port to pin 17 on the L297 for direction of the stepper motor and pin 18 on the L297 for the number of steps the motor will take in that direction. The L297 then sends signals to the L298 in the sequence in which the windings must be powered up to accomplish the task. Then the L298 provides power to the motor windings in the proper order. You will note that motor power is supplied only to the L298 for this purpose. But both chips require +5 volts to function. The eight FR304 diodes clamp the stepper motor windings to motor voltage and ground. Diodes used for this purpose must be fast recovery but could be a different value based on the amperage needed by the stepper motors used. This protects the L298 from the induced high voltages generated by the stepper motor when the any of the windings are turned off.

Pins 1 and 15 on the L298 are connected to two 1/2-ohm power resistors connected to ground. All drive currents used by the stepper's two field windings are passed through these resistors. The resistor connected to pin 1 takes the current from one of the two field windings, while the pin 15 takes the current from the other field winding. These two resistors give the controlling L297 a method of measuring the current being induced within the motor. The L297 measures the voltage drop across these resistors to control the PWM chopper circuit used to control the current within the windings of the stepper motor. The 2.2K and 10K resistors connected to pin 15 (Vref) on the L297 are used to set up a voltage divider. The resulting voltage applied to the Vref pin is used as a set point against the measured voltage coming from the field windings. When the set point is reached, the power drive stage within the L298 driving that winding is turned off, allowing the FR304 diodes to discharge the field winding. The field winding stays off until the internal oscillator within the L297 times out and turns the field winding back on. The 22K resistor and the 3.3nF capacitor connected to pin 16 on the L297 set up the timer's chopper rate. C2, C3, and C4 filter the power supply for the electronics and motors. JP1 provides power and ground for the circuit and the motors. JP2 is jumped to include the circuit connected to pin 1 with pin 16 on the L297 on only one board of the three used; the other two have pin 16 on the L297 jumped to ground. JP3 pin 1 accepts step signals; pin 2 accepts direction signals; pin 3 is used to connect all driver boards that need to be synced and pin 4 is ground. JP4 is set to bring pin 19 on the L297 high at pin 1 to provide full step motor drive or low at pin 3 for half step. JP5 is set to bring pin 11 on the L297 high at pin 1 to use phase driving or low at pin 3 to inhibit drive.

Don't worry too much about understanding how this circuit works; as long as you follow the directions closely, the board you build will function properly.

The Interface Board

This board is really only a gateway that allows the computer to send and receive signals to the drivers and limit switches. There isn't much to it aside from a connector for the straight-through parallel port cable and some connections for the wires coming from each of the driver boards and limit switches. If you refer to the schematic (**Figure 2.7**) you will see why the parts list is short.

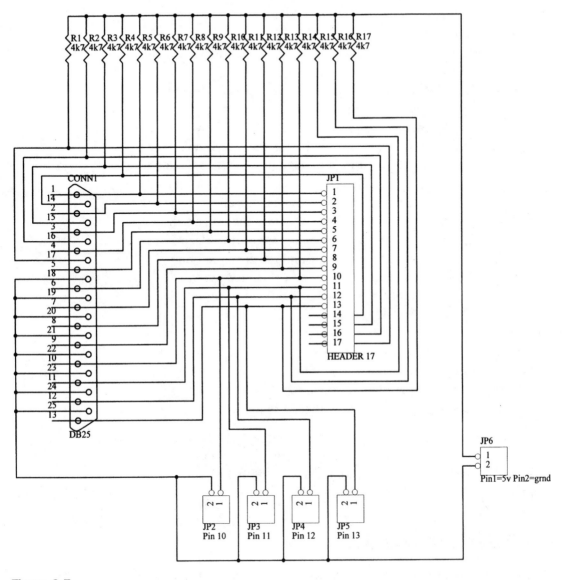

Figure 2.7

Interface schematic.

Part	Quantity	Description	Table 2.2
Conn 1	1	DB 25 connector	Interface Components List
R1-17	17	4.7K ohm resistor	
JP1-5	1-17 pin, 5-2 pin	Header material	

The 4.7K resistors limit current to protect the circuit and the parallel port. Better methods of protecting the parallel port are to use an optical isolation circuit or install a parallel port card specifically to be used with the interface; the cards are cheap insurance against damaging a motherboard. I bought a used computer solely for use with these boards because if I destroy it I won't lose years of accumulated files.

The pin out of the interface board is straightforward. Pins 1–17 on JP1 are connected to pins 1–17 of the parallel port of your computer. JP2 is connected to pin 10 on JP1 at pin 1 and to ground at pin 2. JP3 is connected to pin 11 on JP1 at pin 1 and to ground at pin 2. JP4 is connected to pin 12 on JP1 and to ground at pin 2. JP5 is connected to pin 13 on JP1 and pin 2 is connected to ground. I brought pins 10, 11, 12, and 13 on JP1 out to separate connectors to make hooking them to limit switches a little easier. At JP6, pin 1 is to be connected to 5 volts and pin 2 to ground on the power supply.

This chapter will have given you an understanding of the circuits that are needed to connect and control the stepper motors with a computer. You will also have become familiar with the integrated circuits that the drivers are built around. This understanding will enable you to better troubleshoot your boards when they are complete. The next chapter deals with making the printed circuit boards using the toner transfer method.

SGS-THOMSON
MICROELECTRONICS

L297
L297D

STEPPER MOTOR CONTROLLERS

- NORMAL/WAWE DRIVE
- HALF/FULL STEP MODES
- CLOCKWISE/ANTICLOCKWISE DIRECTION
- SWITCHMODE LOAD CURRENT REGULA-TION
- PROGRAMMABLE LOAD CURRENT
- FEW EXTERNAL COMPONENTS
- RESET INPUT & HOME OUTPUT
- ENABLE INPUT

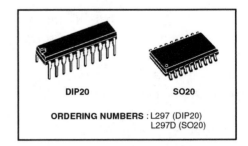

DIP20　　**SO20**

ORDERING NUMBERS : L297 (DIP20)
L297D (SO20)

DESCRIPTION

The L297/A/D Stepper Motor Controller IC gener-ates four phase drive signals for two phase bipolar and four phase unipolar step motors in microcom-puter-controlled applications. The motor can be driven in half step, normal and wawe drive modes and on-chip PWM chopper circuits permit switch-mode control of the current in the windings. A feature of this device is that it requires only clock, direction and mode input signals. Since the phase are generated internally the burden on the micro-processor, and the programmer, is greatly reduced. Mounted in DIP20 and SO20 packages, the L297 can be used with monolithic bridge drives such as the L298N or L293E, or with discrete transistors and darlingtons.

ABSOLUTE MAXIMUM RATINGS

Symbol	Parameter	Value	Unit
V_s	Supply voltage	10	V
V_i	Input signals	7	V
P_{tot}	Total power dissipation ($T_{amb} = 70°C$)	1	W
T_{stg}, T_j	Storage and junction temperature	-40 to + 150	°C

TWO PHASE BIPOLAR STEPPER MOTOR CONTROL CIRCUIT

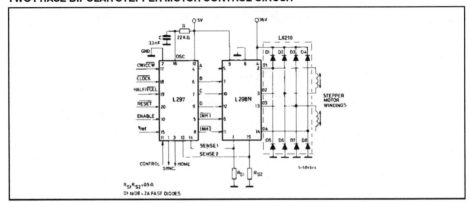

L297-L297D

PIN CONNECTION (Top view)

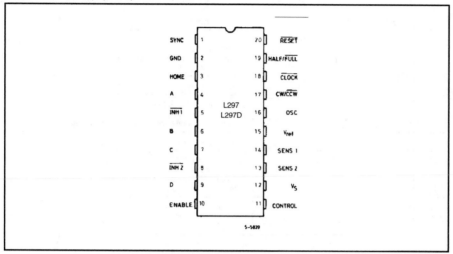

BLOCK DIAGRAM (L297/L297D)

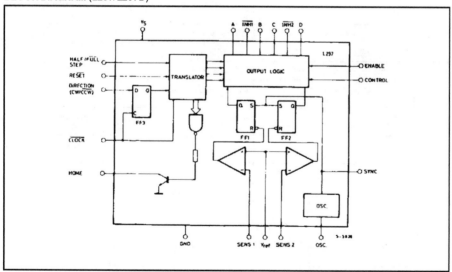

SGS-THOMSON
MICROELECTRONICS

23

PIN FUNCTIONS - L297/L297D

N°	NAME	FUNCTION
1	SYNC	Output of the on-chip chopper oscillator. The SYNC connections The SYNC connections of all L297s to be synchronized are connected together and the oscillator components are omitted on all but one. If an external clock source is used it is injected at this terminal.
2	GND	Ground connection.
3	HOME	Open collector output that indicates when the L297 is in its initial state (ABCD = 0101). The transistor is open when this signal is active.
4	A	Motor phase A drive signal for power stage.
5	$\overline{INH1}$	Active low inhibit control for driver stage of A and B phases. When a bipolar bridge is used this signal can be used to ensure fast decay of load current when a winding is de-energized. Also used by chopper to regulate load current if CONTROL input is low.
6	B	Motor phase B drive signal for power stage.
7	C	Motor phase C drive signal for power stage.
8	$\overline{INH2}$	Active low inhibit control for drive stages of C and D phases. Same functions as INH1.
9	D	Motor phase D drive signal for power stage.
10	ENABLE	Chip enable input. When low (inactive) INH1, INH2, A, B, C and D are brought low.
11	CONTROL	Control input that defines action of chopper. When low chopper acts on INH1 and INH2; when high chopper acts on phase lines ABCD.
12	V_s	5V supply input.
13	$SENS_2$	Input for load current sense voltage from power stages of phases C and D.
14	$SENS_1$	Input for load current sense voltage from power stages of phases A and B.
15	V_{ref}	Reference voltage for chopper circuit. A voltage applied to this pin determines the peak load current.
16	OSC	An RC network (R to V_{CC}, C to ground) connected to this terminal determines the chopper rate. This terminal is connected to ground on all but one device in synchronized multi - L297 configurations. $f \cong 1/0.69\,RC$
17	CW/$\overline{CCW}$	Clockwise/counterclockwise direction control input. Physical direction of motor rotation also depends on connection of windings. Synchronized internally therefore direction can be changed at any time.
18	$\overline{CLOCK}$	Step clock. An active low pulse on this input advances the motor one increment. The step occurs on the rising edge of this signal.

L297-L297D

PIN FUNCTIONS - L297/L297D (continued)

N°	NAME	FUNCTION
19	HALF/FULL	Half/full step select input. When high selects half step operation, when low selects full step operation. One-phase-on full step mode is obtained by selecting FULL when the L297's translator is at an even-numbered state. Two-phase-on full step mode is set by selecting FULL when the translator is at an odd numbered position. (The home position is designate state 1).
20	RESET	Reset input. An active low pulse on this input restores the translator to the home position (state 1, ABCD = 0101).

THERMAL DATA

Symbol	Parameter		DIP20	SO20	Unit
$R_{th-j-amb}$	Thermal resistance junction-ambient	max	80	100	°C/W

CIRCUIT OPERATION

The L297 is intended for use with a dual bridge driver, quad darlington array or discrete power devices in step motor driving applications. It receives step clock, direction and mode signals from the systems controller (usually a microcomputer chip) and generates control signals for the power stage.

The principal functions are a translator, which generates the motor phase sequences, and a dual PWM chopper circuit which regulates the current in the motor windings. The translator generates three different sequences, selected by the HALF/FULL input. These are normal (two phases energised), wave drive (one phase energised) and half-step (alternately one phase energised/two phases energised). Two inhibit signals are also generated by the L297 in half step and wave drive modes. These signals, which connect directly to the L298's enable inputs, are intended to speed current decay when a winding is de-energised. When the L297 is used to drive a unipolar motor the chopper acts on these lines.

An input called CONTROL determines whether the chopper will act on the phase lines ABCD or the inhibit lines INH1 and INH2. When the phase lines are chopped the non-active phase line of each pair (AB or CD) is activated (rather than interrupting the line then active). In L297 + L298 configurations this technique reduces dissipation in the load current sense resistors.

A common on-chip oscillator drives the dual chopper. It supplies pulses at the chopper rate which set the two flip-flops FF1 and FF2. When the current in a winding reaches the programmed peak value the voltage across the sense resistor (connected to one of the sense inputs $SENS_1$ or $SENS_2$) equals V_{ref} and the corresponding comparator resets its flip flop, interrupting the drive current until the next oscillator pulse arrives. The peak current for both windings is programmed by a voltage divider on the V_{ref} input.

Ground noise problems in multiple configurations can be avoided by synchronising the chopper oscillators. This is done by connecting all the SYNC pins together, mounting the oscillator RC network on one device only and grounding the OSC pin on all other devices.

SGS-THOMSON
MICROELECTRONICS

MOTOR DRIVING PHASE SEQUENCES

The L297's translator generates phase sequences for normal drive, wave drive and half step modes. The state sequences and output waveforms for these three modes are shown below. In all cases the translator advances on the low to high transistion of CLOCK.

Clockwise rotation is indicate; for anticlockwise rotation the sequences are simply reversed RESET restores the translator to state 1, where ABCD = 0101.

HALF STEP MODE

Half step mode is selected by a high level on the HALF/FULL input.

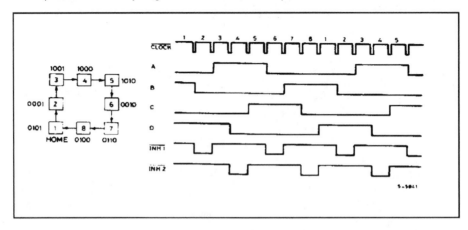

NORMAL DRIVE MODE

Normal drive mode (also called "two-phase-on" drive) is selected by a low level on the HALF/FULL input when the translator is at an odd numbered state (1, 3, 5 or 7). In this mode the INH1 and INH2 outputs remain high throughout.

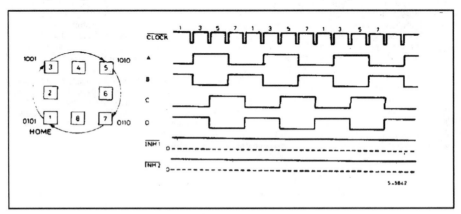

L297-L297D

MOTOR DRIVING PHASE SEQUENCES (continued)

WAVE DRIVE MODE
Wave drive mode (also called "one-phase-on" drive) is selected by a low level on the HALF/FULL input when the translator is at an even numbered state (2, 4, 6 or 8).

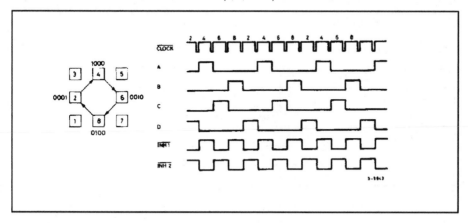

ELECTRICAL CHARACTERISTICS (Refer to the block diagram $T_{amb} = 25°C$, $V_s = 5V$ unless otherwise specified)

Symbol	Parameter	Test conditions		Min.	Typ	Max.	Unit
V_s	Supply voltage (pin 12)			4.75		7	V
I_s	Quiescent supply current (pin 12)	Outputs floating			50	80	mA
V_i	Input voltage (pin 11, 17, 18, 19, 20)	Low				0.6	V
		High		2		V_s	V
I_i	Input current (pin 11, 17, 18, 19, 20)	$V_i = L$			100		μA
		$V_i = H$				10	μA
V_{en}	Enable input voltage (pin 10)	Low				1.3	V
		High		2		V_s	V
I_{en}	Enable input current (pin 10)	$V_{en} = L$				100	μA
		$V_{en} = H$				10	μA
V_o	Phase output voltage (pins 4, 6, 7, 9)	$I_o = 10mA$	V_{OL}			0.4	V
		$I_o = 5mA$	V_{OH}	3.9			V
V_{inh}	Inhibit output voltage (pins 5, 8)	$I_o = 10mA$	$V_{inh\,L}$			0.4	V
		$I_o = 5mA$	$V_{inh\,H}$	3.9			V
V_{SYNC}	Sync Output Voltage	$I_o = 5mA$	$V_{SYNC\,H}$	3.3			V
		$I_o = 5mA$	$V_{SYNC\,V}$			0.8	V

SGS-THOMSON
MICROELECTRONICS

27

ELECTRICAL CHARACTERISTICS (continued)

Symbol	Parameter	Test conditions	Min.	Typ	Max.	Unit
I_{leak}	Leakage current (pin 3)	$V_{CE} = 7$ V			1	µA
V_{sat}	Saturation voltage (pin 3)	I = 5 mA			0.4	V
V_{off}	Comparators offset voltage (pins 13, 14, 15)	$V_{ref} = 1$ V			5	mV
I_o	Comparator bias current (pins 13, 14, 15)		-100		10	µA
V_{ref}	Input reference voltage (pin 15)		0		3	V
t_{CLK}	Clock time		0.5			µs
t_S	Set up time		1			µs
t_H	Hold time		4			µs
t_R	Reset time		1			µs
t_{RCLK}	Reset to clock delay		1			µs

Figure 1.

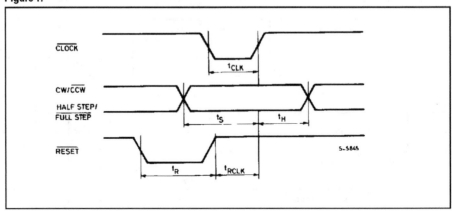

L297-L297D

APPLICATION INFORMATION

TWO PHASE BIPOLAR STEPPER MOTOR CONTROL CIRCUIT

This circuit drives bipolar stepper motors with winding currents up to 2A. The diodes are fast 2A types.

Figure 2.

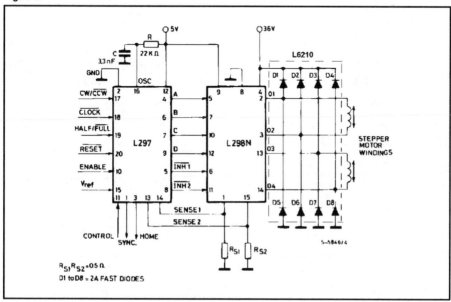

Figure 3 : Synchronising L297s

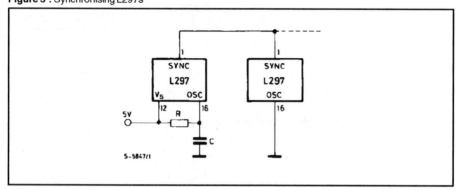

SGS-THOMSON
MICROELECTRONICS

DIP20 PACKAGE MECHANICAL DATA

DIM.	mm			inch		
	MIN.	TYP.	MAX.	MIN.	TYP.	MAX.
a1	0.254			0.010		
B	1.39		1.65	0.055		0.065
b		0.45			0.018	
b1		0.25			0.010	
D			25.4			1.000
E		8.5			0.335	
e		2.54			0.100	
e3		22.86			0.900	
F			7.1			0.280
I			3.93			0.155
L		3.3			0.130	
Z			1.34			0.053

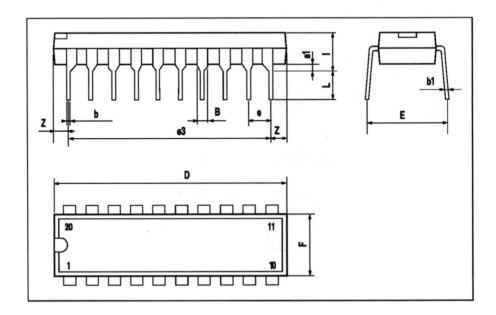

L297-L297D

SO20 PACKAGE MECHANICAL DATA

DIM.	mm			inch		
	MIN.	TYP.	MAX.	MIN.	TYP.	MAX.
A			2.65			0.104
a1	0.1		0.3	0.004		0.012
a2			2.45			0.096
b	0.35		0.49	0.014		0.019
b1	0.23		0.32	0.009		0.013
C		0.5			0.020	
c1	45 (typ.)					
D	12.6		13.0	0.496		0.512
E	10		10.65	0.394		0.419
e		1.27			0.050	
e3		11.43			0.450	
F	7.4		7.6	0.291		0.299
L	0.5		1.27	0.020		0.050
M			0.75			0.030
S	8 (max.)					

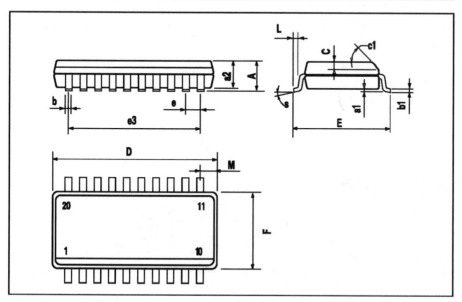

SGS-THOMSON
MICROELECTRONICS

L297-L297D

L298

DUAL FULL-BRIDGE DRIVER

- OPERATING SUPPLY VOLTAGE UP TO 46 V
- TOTAL DC CURRENT UP TO 4 A
- LOW SATURATION VOLTAGE
- OVERTEMPERATURE PROTECTION
- LOGICAL "0" INPUT VOLTAGE UP TO 1.5 V (HIGH NOISE IMMUNITY)

DESCRIPTION

The L298 is an integrated monolithic circuit in a 15-lead Multiwatt and PowerSO20 packages. It is a high voltage, high current dual full-bridge driver designed to accept standard TTL logic levels and drive inductive loads such as relays, solenoids, DC and stepping motors. Two enable inputs are provided to enable or disable the device independently of the input signals. The emitters of the lower transistors of each bridge are connected together and the corresponding external terminal can be used for the con-

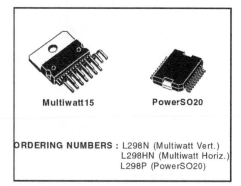

Multiwatt15   **PowerSO20**

ORDERING NUMBERS : L298N (Multiwatt Vert.)
L298HN (Multiwatt Horiz.)
L298P (PowerSO20)

nection of an external sensing resistor. An additional supply input is provided so that the logic works at a lower voltage.

BLOCK DIAGRAM

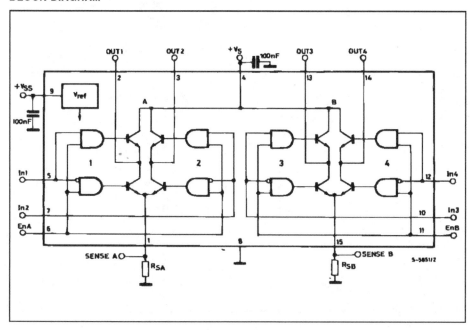

L298

ABSOLUTE MAXIMUM RATINGS

Symbol	Parameter	Value	Unit
V_S	Power Supply	50	V
V_{SS}	Logic Supply Voltage	7	V
V_I, V_{en}	Input and Enable Voltage	-0.3 to 7	V
I_O	Peak Output Current (each Channel) – Non Repetitive (t = 100µs) –Repetitive (80% on –20% off; t_{on} = 10ms) –DC Operation	 3 2.5 2	 A A A
V_{sens}	Sensing Voltage	-1 to 2.3	V
P_{tot}	Total Power Dissipation (T_{case} = 75°C)	25	W
T_{op}	Junction Operating Temperature	-25 to 130	°C
T_{stg}, T_j	Storage and Junction Temperature	-40 to 150	°C

PIN CONNECTIONS (top view)

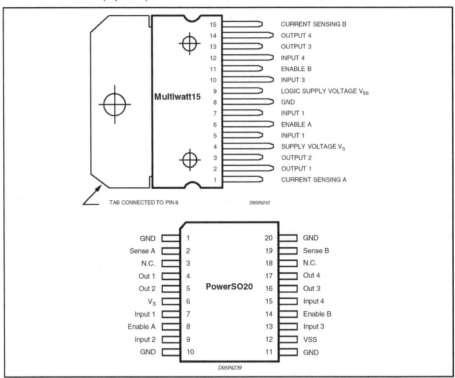

Multiwatt15

15 CURRENT SENSING B
14 OUTPUT 4
13 OUTPUT 3
12 INPUT 4
11 ENABLE B
10 INPUT 3
9 LOGIC SUPPLY VOLTAGE V_{SS}
8 GND
7 INPUT 1
6 ENABLE A
5 INPUT 1
4 SUPPLY VOLTAGE V_S
3 OUTPUT 2
2 OUTPUT 1
1 CURRENT SENSING A

TAB CONNECTED TO PIN 8   D95IN240

PowerSO20

GND 1	20 GND
Sense A 2	19 Sense B
N.C. 3	18 N.C.
Out 1 4	17 Out 4
Out 2 5	16 Out 3
V_S 6	15 Input 4
Input 1 7	14 Enable B
Enable A 8	13 Input 3
Input 2 9	12 VSS
GND 10	11 GND

D95IN239

THERMAL DATA

Symbol	Parameter		PowerSO20	Multiwatt15	Unit
$R_{th\ j\text{-}case}$	Thermal Resistance Junction-case	Max.	–	3	°C/W
$R_{th\ j\text{-}amb}$	Thermal Resistance Junction-ambient	Max.	13 (*)	35	°C/W

(*) Mounted on aluminum substrate

PIN FUNCTIONS (refer to the block diagram)

MW.15	PowerSO	Name	Function
1;15	2;19	Sense A; Sense B	Between this pin and ground is connected the sense resistor to control the current of the load.
2;3	4;5	Out 1; Out 2	Outputs of the Bridge A; the current that flows through the load connected between these two pins is monitored at pin 1.
4	6	V_S	Supply Voltage for the Power Output Stages. A non-inductive 100nF capacitor must be connected between this pin and ground.
5;7	7;9	Input 1; Input 2	TTL Compatible Inputs of the Bridge A.
6;11	8;14	Enable A; Enable B	TTL Compatible Enable Input: the L state disables the bridge A (enable A) and/or the bridge B (enable B).
8	1,10,11,20	GND	Ground.
9	12	VSS	Supply Voltage for the Logic Blocks. A 100nF capacitor must be connected between this pin and ground.
10; 12	13;15	Input 3; Input 4	TTL Compatible Inputs of the Bridge B.
13; 14	16;17	Out 3; Out 4	Outputs of the Bridge B. The current that flows through the load connected between these two pins is monitored at pin 15.
—	3;18	N.C.	Not Connected

ELECTRICAL CHARACTERISTICS (V_S = 42V; V_{SS} = 5V, T_j = 25°C; unless otherwise specified)

Symbol	Parameter	Test Conditions		Min.	Typ.	Max.	Unit
V_S	Supply Voltage (pin 4)	Operative Condition		V_{IH} +2.5		46	V
V_{SS}	Logic Supply Voltage (pin 9)			4.5	5	7	V
I_S	Quiescent Supply Current (pin 4)	V_{en} = H; I_L = 0	V_i = L		13	22	mA
			V_i = H		50	70	mA
		V_{en} = L	V_i = X			4	mA
I_{SS}	Quiescent Current from V_{SS} (pin 9)	V_{en} = H; I_L = 0	V_i = L		24	36	mA
			V_i = H		7	12	mA
		V_{en} = L	V_i = X			6	mA
V_{iL}	Input Low Voltage (pins 5, 7, 10, 12)			−0.3		1.5	V
V_{iH}	Input High Voltage (pins 5, 7, 10, 12)			2.3		VSS	V
I_{iL}	Low Voltage Input Current (pins 5, 7, 10, 12)	V_i = L				−10	µA
I_{iH}	High Voltage Input Current (pins 5, 7, 10, 12)	Vi = H ≤ V_{SS} −0.6V			30	100	µA
V_{en} = L	Enable Low Voltage (pins 6, 11)			−0.3		1.5	V
V_{en} = H	Enable High Voltage (pins 6, 11)			2.3		V_{SS}	V
I_{en} = L	Low Voltage Enable Current (pins 6, 11)	V_{en} = L				−10	µA
I_{en} = H	High Voltage Enable Current (pins 6, 11)	V_{en} = H ≤ V_{SS} −0.6V			30	100	µA
$V_{CEsat(H)}$	Source Saturation Voltage	I_L = 1A		0.95	1.35	1.7	V
		I_L = 2A			2	2.7	V
$V_{CEsat(L)}$	Sink Saturation Voltage	I_L = 1A (5)		0.85	1.2	1.6	V
		I_L = 2A (5)			1.7	2.3	V
V_{CEsat}	Total Drop	I_L = 1A (5)		1.80		3.2	V
		I_L = 2A (5)				4.9	V
V_{sens}	Sensing Voltage (pins 1, 15)			−1 (1)		2	V

ST

L298

ELECTRICAL CHARACTERISTICS (continued)

Symbol	Parameter	Test Conditions	Min.	Typ.	Max.	Unit
$T_1 (V_i)$	Source Current Turn-off Delay	$0.5\ V_i$ to $0.9\ I_L$ (2); (4)		1.5		μs
$T_2 (V_i)$	Source Current Fall Time	$0.9\ I_L$ to $0.1\ I_L$ (2); (4)		0.2		μs
$T_3 (V_i)$	Source Current Turn-on Delay	$0.5\ V_i$ to $0.1\ I_L$ (2); (4)		2		μs
$T_4 (V_i)$	Source Current Rise Time	$0.1\ I_L$ to $0.9\ I_L$ (2); (4)		0.7		μs
$T_5 (V_i)$	Sink Current Turn-off Delay	$0.5\ V_i$ to $0.9\ I_L$ (3); (4)		0.7		μs
$T_6 (V_i)$	Sink Current Fall Time	$0.9\ I_L$ to $0.1\ I_L$ (3); (4)		0.25		μs
$T_7 (V_i)$	Sink Current Turn-on Delay	$0.5\ V_i$ to $0.9\ I_L$ (3); (4)		1.6		μs
$T_8 (V_i)$	Sink Current Rise Time	$0.1\ I_L$ to $0.9\ I_L$ (3); (4)		0.2		μs
fc (V_i)	Commutation Frequency	$I_L = 2A$		25	40	KHz
$T_1 (V_{en})$	Source Current Turn-off Delay	$0.5\ V_{en}$ to $0.9\ I_L$ (2); (4)		3		μs
$T_2 (V_{en})$	Source Current Fall Time	$0.9\ I_L$ to $0.1\ I_L$ (2); (4)		1		μs
$T_3 (V_{en})$	Source Current Turn-on Delay	$0.5\ V_{en}$ to $0.1\ I_L$ (2); (4)		0.3		μs
$T_4 (V_{en})$	Source Current Rise Time	$0.1\ I_L$ to $0.9\ I_L$ (2); (4)		0.4		μs
$T_5 (V_{en})$	Sink Current Turn-off Delay	$0.5\ V_{en}$ to $0.9\ I_L$ (3); (4)		2.2		μs
$T_6 (V_{en})$	Sink Current Fall Time	$0.9\ I_L$ to $0.1\ I_L$ (3); (4)		0.35		μs
$T_7 (V_{en})$	Sink Current Turn-on Delay	$0.5\ V_{en}$ to $0.9\ I_L$ (3); (4)		0.25		μs
$T_8 (V_{en})$	Sink Current Rise Time	$0.1\ I_L$ to $0.9\ I_L$ (3); (4)		0.1		μs

1) 1)Sensing voltage can be $-1\ V$ for $t \leq 50\ \mu sec$; in steady state V_{sens} min $\geq -0.5\ V$.
2) See fig. 2.
3) See fig. 4.
4) The load must be a pure resistor.

Figure 1 : Typical Saturation Voltage vs. Output Current.

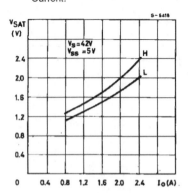

Figure 2 : Switching Times Test Circuits.

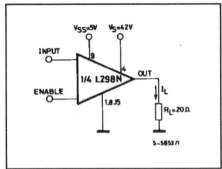

Note : For INPUT Switching, set EN = H
For ENABLE switching, set IN = H

Figure 3 : Source Current Delay Times vs. Input or Enable Switching.

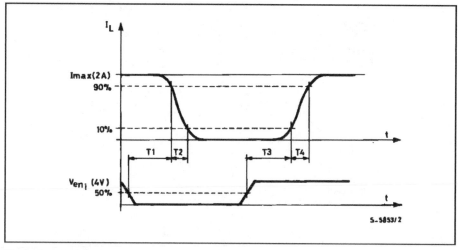

Figure 4 : Switching Times Test Circuits.

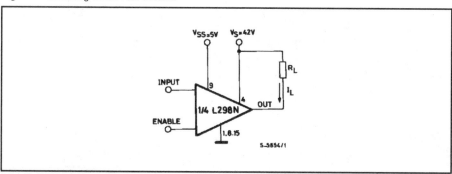

Note : For INPUT Switching, set EN = H
For ENABLE Switching, set IN = L

L298

Figure 5 : Sink Current Delay Times vs. Input 0 V Enable Switching.

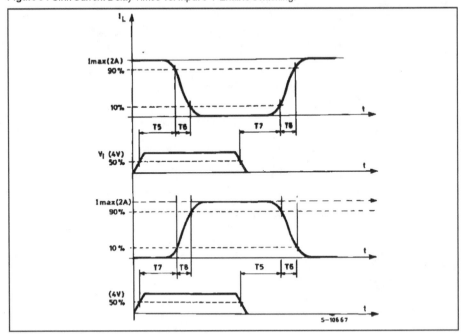

Figure 6 : Bidirectional DC Motor Control.

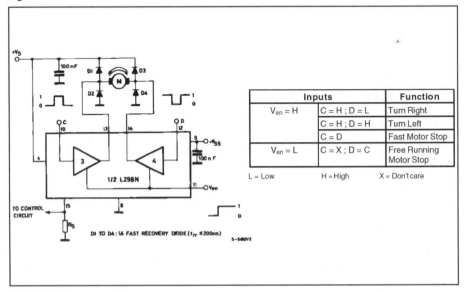

Inputs		Function
$V_{en} = H$	C = H ; D = L	Turn Right
	C = H ; D = H	Turn Left
	C = D	Fast Motor Stop
$V_{en} = L$	C = X ; D = C	Free Running Motor Stop

L = Low   H = High   X = Don't care

Figure 7 : For higher currents, outputs can be paralleled. Take care to parallel channel 1 with channel 4 and channel 2 with channel 3.

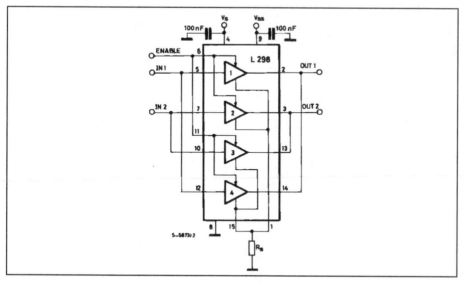

APPLICATION INFORMATION (Refer to the block diagram)

1.1. POWER OUTPUT STAGE

The L298 integrates two power output stages (A ; B). The power output stage is a bridge configuration and its outputs can drive an inductive load in common or differenzial mode, depending on the state of the inputs. The current that flows through the load comes out from the bridge at the sense output : an external resistor (R_{SA} ; R_{SB}.) allows to detect the intensity of this current.

1.2. INPUT STAGE

Each bridge is driven by means of four gates the input of which are In1 ; In2 ; EnA and In3 ; In4 ; EnB. The In inputs set the bridge state when The En input is high ; a low state of the En input inhibits the bridge. All the inputs are TTL compatible.

2. SUGGESTIONS

A non inductive capacitor, usually of 100 nF, must be foreseen between both Vs and Vss, to ground, as near as possible to GND pin. When the large capacitor of the power supply is too far from the IC, a second smaller one must be foreseen near the L298.

The sense resistor, not of a wire wound type, must be grounded near the negative pole of Vs that must be near the GND pin of the I.C.

Each input must be connected to the source of the driving signals by means of a very short path.

Turn-On and Turn-Off : Before to Turn-ON the Supply Voltage and before to Turn it OFF, the Enable input must be driven to the Low state.

3. APPLICATIONS

Fig 6 shows a bidirectional DC motor control Schematic Diagram for which only one bridge is needed. The external bridge of diodes D1 to D4 is made by four fast recovery elements (trr ≤ 200 nsec) that must be chosen of a VF as low as possible at the worst case of the load current.

The sense output voltage can be used to control the current amplitude by chopping the inputs, or to provide overcurrent protection by switching low the enable input.

The brake function (Fast motor stop) requires that the Absolute Maximum Rating of 2 Amps must never be overcome.

When the repetitive peak current needed from the load is higher than 2 Amps, a paralleled configuration can be chosen (See Fig.7).

An external bridge of diodes are required when inductive loads are driven and when the inputs of the IC are chopped; Shottky diodes would be preferred.

L298

This solution can drive until 3 Amps In DC operation and until 3.5 Amps of a repetitive peak current.

On Fig 8 it is shown the driving of a two phase bipolar stepper motor ; the needed signals to drive the inputs of the L298 are generated, in this example, from the IC L297.

Fig 9 shows an example of P.C.B. designed for the application of Fig 8.

Fig 10 shows a second two phase bipolar stepper motor control circuit where the current is controlled by the I.C. L6506.

Figure 8 : Two Phase Bipolar Stepper Motor Circuit.

This circuit drives bipolar stepper motors with winding currents up to 2 A. The diodes are fast 2 A types.

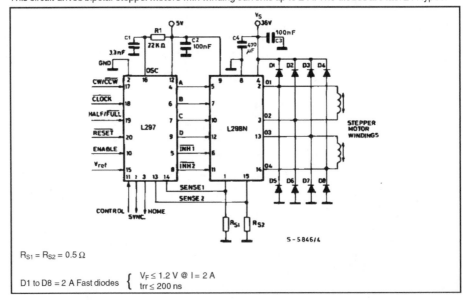

$R_{S1} = R_{S2} = 0.5 \, \Omega$

D1 to D8 = 2 A Fast diodes $\begin{cases} V_F \leq 1.2 \text{ V @ I = 2 A} \\ trr \leq 200 \text{ ns} \end{cases}$

Figure 9 : Suggested Printed Circuit Board Layout for the Circuit of fig. 8 (1:1 scale).

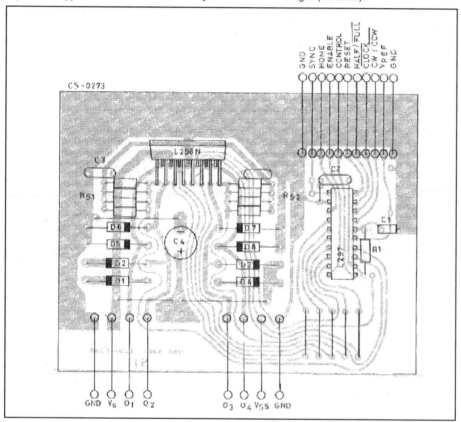

Figure 10 : Two Phase Bipolar Stepper Motor Control Circuit by Using the Current Controller L6506.

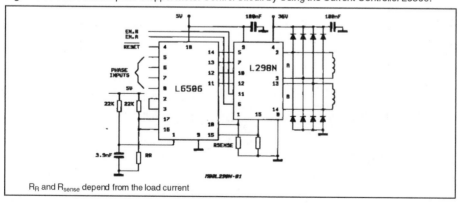

R_R and R_{sense} depend from the load current

L298

DIM.	mm			inch		
	MIN.	TYP.	MAX.	MIN.	TYP.	MAX.
A			5			0.197
B			2.65			0.104
C			1.6			0.063
D		1			0.039	
E	0.49		0.55	0.019		0.022
F	0.66		0.75	0.026		0.030
G	1.02	1.27	1.52	0.040	0.050	0.060
G1	17.53	17.78	18.03	0.690	0.700	0.710
H1	19.6			0.772		
H2			20.2			0.795
L	21.9	22.2	22.5	0.862	0.874	0.886
L1	21.7	22.1	22.5	0.854	0.870	0.886
L2	17.65		18.1	0.695		0.713
L3	17.25	17.5	17.75	0.679	0.689	0.699
L4	10.3	10.7	10.9	0.406	0.421	0.429
L7	2.65		2.9	0.104		0.114
M	4.25	4.55	4.85	0.167	0.179	0.191
M1	4.63	5.08	5.53	0.182	0.200	0.218
S	1.9		2.6	0.075		0.102
S1	1.9		2.6	0.075		0.102
Dia1	3.65		3.85	0.144		0.152

OUTLINE AND MECHANICAL DATA

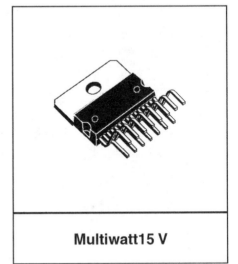

Multiwatt15 V

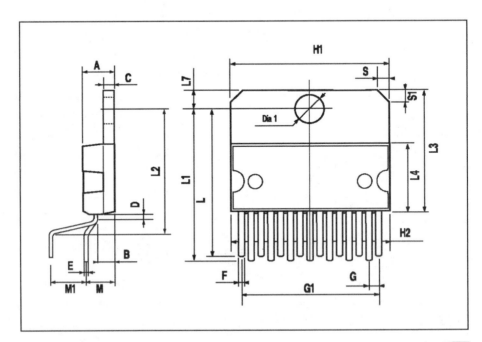

L298

DIM.	mm			inch		
	MIN.	TYP.	MAX.	MIN.	TYP.	MAX.
A			5			0.197
B			2.65			0.104
C			1.6			0.063
E	0.49		0.55	0.019		0.022
F	0.66		0.75	0.026		0.030
G	1.14	1.27	1.4	0.045	0.050	0.055
G1	17.57	17.78	17.91	0.692	0.700	0.705
H1	19.6			0.772		
H2			20.2			0.795
L		20.57			0.810	
L1		18.03			0.710	
L2		2.54			0.100	
L3	17.25	17.5	17.75	0.679	0.689	0.699
L4	10.3	10.7	10.9	0.406	0.421	0.429
L5		5.28			0.208	
L6		2.38			0.094	
L7	2.65		2.9	0.104		0.114
S	1.9		2.6	0.075		0.102
S1	1.9		2.6	0.075		0.102
Dia1	3.65		3.85	0.144		0.152

OUTLINE AND MECHANICAL DATA

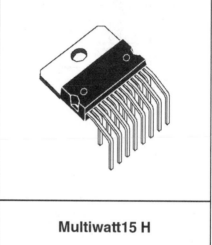

Multiwatt15 H

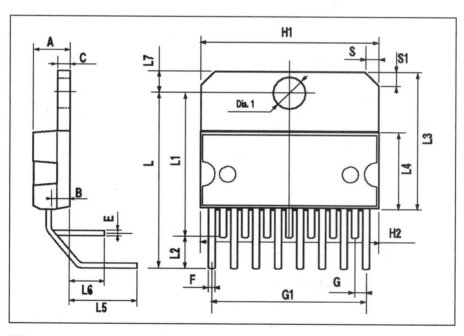

L298

DIM.	mm			inch		
	MIN.	TYP.	MAX.	MIN.	TYP.	MAX.
A			3.6			0.142
a1	0.1		0.3	0.004		0.012
a2			3.3			0.130
a3	0		0.1	0.000		0.004
b	0.4		0.53	0.016		0.021
c	0.23		0.32	0.009		0.013
D (1)	15.8		16	0.622		0.630
D1	9.4		9.8	0.370		0.386
E	13.9		14.5	0.547		0.570
e		1.27			0.050	
e3		11.43			0.450	
E1 (1)	10.9		11.1	0.429		0.437
E2			2.9			0.114
E3	5.8		6.2	0.228		0.244
G	0		0.1	0.000		0.004
H	15.5		15.9	0.610		0.626
h			1.1			0.043
L	0.8		1.1	0.031		0.043
N	10° (max.)					
S	8° (max.)					
T		10			0.394	

(1) "D and F" do not include mold flash or protrusions.
- Mold flash or protrusions shall not exceed 0.15 mm (0.006").
- Critical dimensions: "E", "G" and "a3"

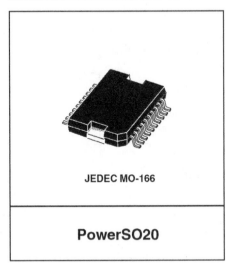

OUTLINE AND MECHANICAL DATA

JEDEC MO-166

PowerSO20

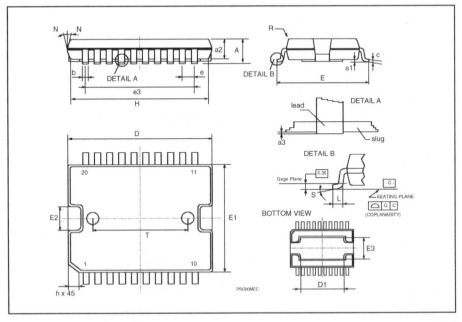

L298

APPLICATION NOTE

THE L297 STEPPER MOTOR CONTROLLER

The L297 integrates all the control circuitry required to control bipolar and unipolar stepper motors. Used with a dual bridge driver such as the L298N forms a complete microprocessor-to-bipolar stepper motor interface. Unipolar stepper motor can be driven with an L297 plus a quad darlington array. This note describes the operation of the circuit and shows how it is used.

The L297 Stepper Motor Controller is primarily intended for use with an L298N or L293E bridge driver in stepper motor driving applications.

It receives control signals from the system's controller, usually a microcomputer chip, and provides all the necessary drive signals for the power stage. Additionally, it includes two PWM chopper circuits to regulate the current in the motor windings.

With a suitable power actuator the L297 drives two phase bipolar permanent magnet motors, four phase unipolar permanent magnet motors and four phase variable reluctance motors. Moreover, it handles normal, wave drive and half step drive modes. (This is all explained in the section "Stepper Motor Basics").

Two versions of the device are available : the regular L297 and a special version called L297A. The L297A incorporates a step pulse doubler and is designed specifically for floppy-disk head positioning applications.

ADVANTAGES

The L297 + driver combination has many advantages : very few components are required (so assembly costs are low, reliability high and little space required), software development is simplified and the burden on the micro is reduced. Further, the choice of a two-chip approach gives a high degree of flexibility-the L298N can be used on its own for DC motors and the L297 can be used with any power stage, including discrete power devices (it provides 20mA drive for this purpose).

Figure 1 : In this typical configuration an L297 stepper motor controller and L298 dual bridge driver combine to form a complete microprocessor to bipolar stepper motor interface.

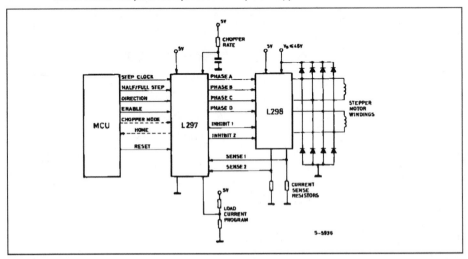

APPLICATION NOTE

For bipolar motors with winding currents up to 2A the L297 should be used with the L298N ; for winding currents up to 1A the L293E is recommended (the L293 will also be useful if the chopper isn't needed). Higher currents are obtained with power transistors or darlingtons and for unipolar motors a darlington array such as the ULN2075B is suggested. The block diagram, figure 1, shows a typical system.

Applications of the L297 can be found almost everywhere ... printers (carriage position, daisy position, paper feed, ribbon feed), typewriters, plotters, numerically controlled machines, robots, floppy disk drives, electronic sewing machines, cash registers, photocopiers, telex machines, electronic carburetos, telecopiers, photographic equipment, paper tape readers, optical character recognisers, electric valves and so on.

The L297 is made with SGS' analog/digital compatible I²L technology (like Zodiac) and is assembled in a 20-pin plastic DIP. A 5V supply is used and all signal lines are TTL/CMOS compatible or open collector transistors. High density is one of the key features of the technology so the L297 die is very compact.

THE L298N AND L293E

Since the L297 is normally used with an L298N or L293E bridge driver a brief review of these devices will make the rest of this note easier to follow.

The L298N and L293E contain two bridge driver stages, each controlled by two TTL-level logic inputs and a TTL-level enable input. In addition, the emitter connections of the lower transistors are brought out to external terminals to allow the connection of current sensing resistors (figure 2).

For the L298N SGS' innovative ion-implanted high voltage/high current technology is used, allowing it to handle effective powers up to 160W (46V supply, 2A per bridge). A separate 5V logic supply input is provided to reduce dissipation and to allow direct connection to the L297 or other control logic.

In this note the pins of the L298N are labelled with the pin names of the corresponding L297 terminals to avoid unnecessary confusion.

The L298N is supplied in a 15-lead Multiwatt plastic power package. It's smaller brother, the functionally identical L293E, is packaged in a Powerdip – a copper frame DIP that uses the four center pins to conduct heat to the circuit board copper.

Figure 2 : The L298N contains two bridge drivers (four push pull stages) each controlled by two logic inputs and an enable input. External emitter connections are provided for current sense resistors. The L293E has external connections for all four emitters.

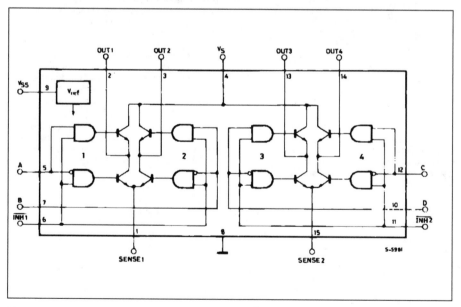

SGS-THOMSON
MICROELECTRONICS

47

STEPPER MOTOR BASICS

There are two basic types of stepper motor in common use : permanent magnet and variable reluctance. Permanent magnet motors are divided into bipolar and unipolar types.

BIPOLAR MOTORS

Simplified to the bare essentials, a bipolar permanent magnet motor consists of a rotating permanent magnet surrounded by stator poles carrying the windings (figure 3). Bidirectional drive current is used and the motor is stepped by switching the windings in sequence.

For a motor of this type there are three possible drive sequences.

Figure 3 : Greatly simplified, a bipolar permanent magnet stepper motor consist of a rotaring magnet surrounded by stator poles as shown.

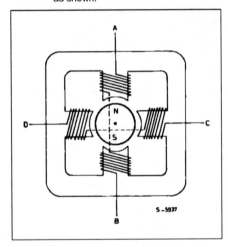

The first is to energize the windings in the sequence AB/CD/BA/DC (BA means that the winding AB is energized but in the opposite sense). This sequence is known as "one phase on" full step or wave drive

mode. Only one phase is energized at any given moment (figure 4a).

The second possibility is to energize both phases together, so that the rotor always aligns itself between two pole positions. Called "two-phase-on" full step, this mode is the normal drive sequence for a bipolar motor and gives the highest torque (figure 4b).

The third option is to energize one phase, then two, then one, etc., so that the motor moves in half step increments. This sequence, known as half step mode, halves the effective step angle of the motor but gives a less regular torque (figure 4c).

For rotation in the opposite direction (counter-clockwise) the same three sequences are used, except of course that the order is reserved.

As shown in these diagrams the motor would have a step angle of 90°. Real motors have multiple poles to reduce the step angle to a few degrees but the number of windings and the drive sequences are unchanged. A typical bipolar stepper motor is shown in figure 5.

UNIPOLAR MOTORS

A unipolar permanent magnet motor is identical to the bipolar machine described above except that bifilar windings are used to reverse the stator flux, rather than bidirectional drive (figure 6).

This motor is driven in exactly the same way as a bipolar motor except that the bridge drivers are replaced by simple unipolar stages - four darlingtons or a quad darlington array. Clearly, unipolar motors are more expensive because thay have twice as many windings. Moreover, unipolar motors give less torque for a given motor size because the windings are made with thinner wire. In the past unipolar motors were attractive to designers because they simplify the driver stage. Now that monolithic push pull drivers like the L298N are available bipolar motors are becoming more popular.

All permanent magnet motors suffer from the counter EMF generated by the rotor, which limits the rotation speed. When very high slewing speeds are necessary a variable reluctance motor is used.

APPLICATION NOTE

Figure 4 : The three drive sequences for a two phase bipolar stepper motor. Clockwise rotation is shown.
Figure 4a : Wave drive (one phase on).

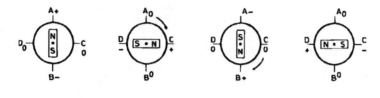

S-5952

Figure 4b : Two phase on drive.

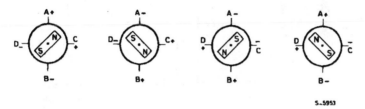

S-5953

Figure 4c : Half step drive.

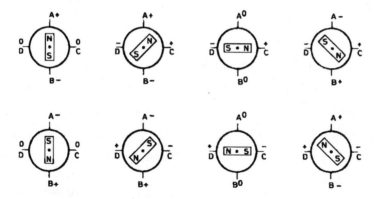

SGS-THOMSON
MICROELECTRONICS

Figure 5 : A real motor. Multiple poles are normally employed to reduce the step angle to a practical value. The principle of operation and drive sequences remain the same.

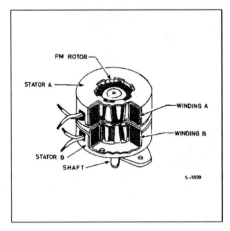

Figure 6 : A unipolar PM motor uses bifilar windings to reverse the flux in each phase.

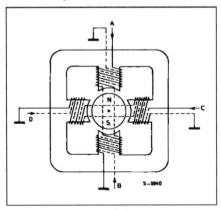

VARIABLE RELUCTANCE MOTORS

A variable reluctance motor has a non-magnetized soft iron rotor with fewer poles than the stator (figure 7). Unipolar drive is used and the motor is stepped by energizing stator pole pairs to align the rotor with the pole pieces of the energized winding.

Once again three different phase sequences can be used. The wave drive sequence is A/C/B/D ; two-

phase-on is AC/CB/BD/DA and the half step sequence is A/AC/C/BC/B/BD/D/DA. Note that the step angle for the motor shown above is 15°, not 45°.

As before, pratical motors normally employ multiple poles to give a much smaller step angle. This does not, however, affect the principle of operation of the drive sequences.

Figure 7 : A variable reluctance motor has a soft iron rotor with fewer poles than the stator. The step angle is 15° for this motor.

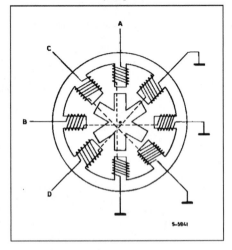

GENERATING THE PHASE SEQUENCES

The heart of the L297 block diagram, figure 8, is a block called the translator which generates suitable phase sequences for half step, one-phase-on full step and two-phase-on full step operation. This block is controlled by two mode inputs – direction (CW/ CCW) and HALF/ FULL – and a step clock which advances the translator from one step to the next.

Four outputs are provided by the translator for subsequent processing by the output logic block which implements the inhibit and chopper functions.

Internally the translator consists of a 3-bit counter plus some combinational logic which generates a basic eight-step gray code sequence as shown in figure 9. All three drive sequences can be generated easily from this master sequence. This state sequence corresponds directly to half step mode, selected by a high level on the HALF/ FULL input.

APPLICATION NOTE

The output waveforms for this sequence are shown in figure 10.

Note that two other signals, $\overline{INH1}$ and $\overline{INH2}$ are generated in this sequence. The purpose of these signals is explained a little further on.

The full step modes are both obtained by skipping alternate states in the eight-step sequence. What happens is that the step clock bypasses the first stage of the 3-bit counter in the translator. The least significant bit of this counter is not affected therefore

the sequence generated depends on the state of the translator when full step mode is selected (the HALF/ $\overline{FULL}$ input brought low).

If full step mode is selected when the translator is at any odd-numbered state we get the two-phase-on full step sequence shown in figure 11.

By contrast, one-phase-on full step mode is obtained by selecting full step mode when the translator is at an even-numbered state (figure 12).

Figure 8 : The L297 contains translator (phase sequence generator), a dual PWM chopper and output control logic.

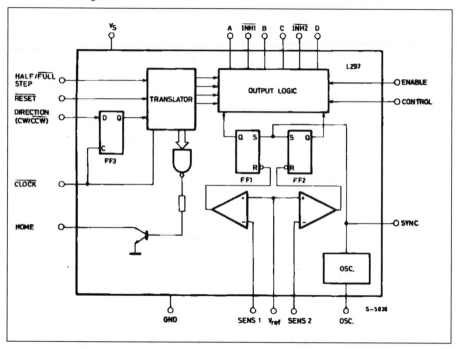

Figure 9 : The eight step master sequence of the translator. This corresponds to half step mode. Clockwise rotation is indicated.

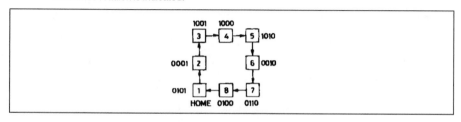

51

Figure 10 : The output waveforms corresponding to the half step sequence. The chopper action in not shown.

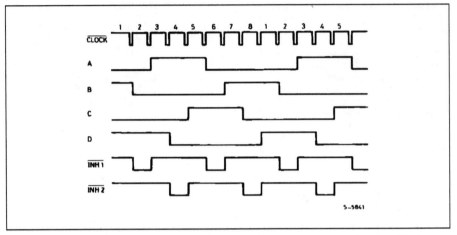

Figure 11 : State sequence and output waveforms for the two phase on sequence. $\overline{INH1}$ and $\overline{INH2}$ remain high throughout.

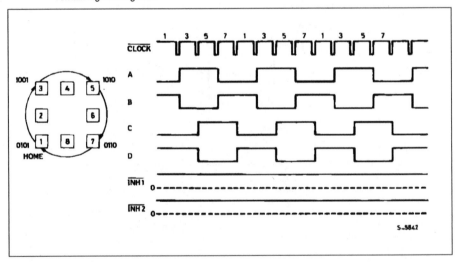

APPLICATION NOTE

Figure 12 : State Sequence and Output Waveforms for Wave Drive (one phase on).

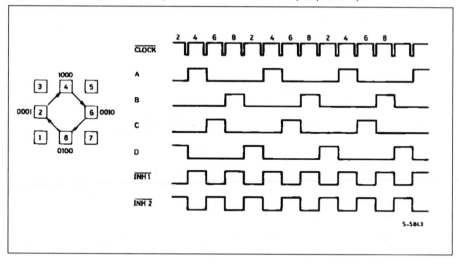

$\overline{\text{INH1}}$ AND $\overline{\text{INH2}}$

In half step and one-phase-on full step modes two other signals are generated: INH1 and INH2. These are inhibit signals which are coupled to the L298N's enable inputs and serve to speed the current decay when a winding is switched off.

Since both windings are energized continuously in two-phase-on full step mode no winding is ever switched off and these signals are not generated.

To see what these signals do let's look at one half of the L298N connected to the first phase of a two-phase bipolar motor (figure 13). Remember that the L298N's A and B inputs determine which transistor in each push pull pair will be on. INH1, on the other hand, turns off all four transistors.

Assume that A is high, B low and current flowing through Q1, Q4 and the motor winding. If A is now brought low the current would recirculate through D2, Q4 and R_s, giving a slow decay and increased dissipation in R_s. If, on a other hand, A is brought low and INH1 is activated, all four transistors are turned off. The current recirculates in this case from ground to V_s via D2 and D3, giving a faster decay thus allowing faster operation of the motor. Also, since the recirculation current does not flow through R_s, a less expensive resistor can be used.

Exactly the same thing happens with the second winding, the other half of the L298 and the signals C, D and INH2.

The $\overline{\text{INH1}}$ and $\overline{\text{INH2}}$ signals are generated by OR functions :

$$A + B = \overline{\text{INH1}} \qquad C + D = \overline{\text{INH2}}$$

However, the output logic is more complex because inhibit lines are also used by the chopper, as we will see further on.

OTHER SIGNALS

Two other signals are connected to the translator block : the RESET input and the HOME output

RESET is an asynchronous reset input which restores the translator block to the home position (state 1, ABCD = 0101). The HOME output (open collector) signals this condition and is intended to the ANDed with the output of a mechanical home position sensor.

Finally, there is an ENABLE input connected to the output logic. A low level on this input brings INH1, INH2, A, B, C and D low. This input is useful to disable the motor driver when the system is initialized.

LOAD CURRENT REGULATION

Some form of load current control is essential to obtain good speed and torque characteristics. There are several ways in which this can be done – switching the supply between two voltages, pulse rate modulation chopping or pulse width modulation chopping.

SGS-THOMSON
MICROELECTRONICS

Figure 13 : When a winding is switched off the inhibit input is activated to speed current decay. If this were not done the current would recirculate through D2 and Q4 in this example. Dissipation in R_s is also reduced.

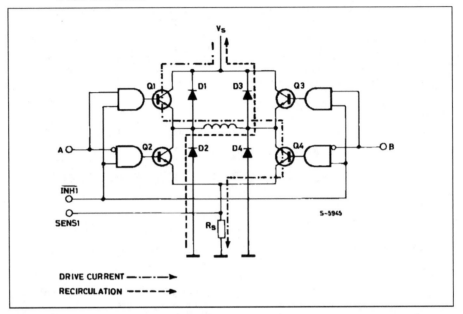

DRIVE CURRENT —·—··—▸

RECIRCULATION ————▸

The L297 provides load current control in the form of two PWM choppers, one for each phase of a bipolar motor or one for each pair of windings for a unipolar motor. (In a unipolar motor the A and B windings are never energized together so thay can share a chopper ; the same applies to C and D).

Each chopper consists of a comparator, a flip flop and an external sensing resistor. A common on chip oscillator supplies pulses at the chopper rate to both choppers.

In each chopper (figure 14) the flip flop is set by each pulse from the oscillator, enabling the output and allowing the load current to increase. As it increases the voltage across the sensing resistor increases, and when this voltage reaches V_{ref} the flip flop is reset, disabling the output until the next oscillator pulse arrives. The output of this circuit (the flip flop's Q output) is therefore a constant rate PWM signal. Note that V_{ref} determines the peak load current.

Figure 14 : Each chopper circuit consists of a comparator, flip flop and external sense resistor. A common oscillator clocks both circuits.

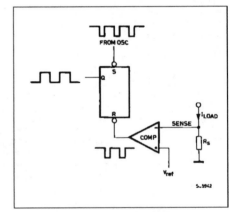

APPLICATION NOTE

PHASE CHOPPING AND INHIBIT CHOPPING

The chopper can act on either the phase lines (ABCD) or on the inhibit lines $\overline{INH1}$ and $\overline{INH2}$. An input named CONTROL decides which. Inhibit chopping is used for unipolar motors but you can choose between phase chopping and inhibit chopping for bipolar motors. The reasons for this choice are best explained with another example.

First let's examine the situation when the phase lines are chopped.

As before, we are driving a two phase bipolar motor and A is high, B low (figure 15). Current therefore flows through Q1, winding, Q4 and R_s. When the voltage across R_s reaches V_{ref} the chopper brings B high to switch off the winding.

The energy stored in the winding is dissipated by current recirculating through Q1 and D3. Current decay through this path is rather slow because the voltage on the winding is low ($V_{CEsat\ Q1} + V_{D3}$) (figure 16).

Why is B pulled high, why push A low ? The reason is to avoid the current decaying through R_s. Since the current recirculates in the upper half of the bridge, current only flows in the sensing resistor when the winding is driven. Less power is therefore dissipated in R_s and we can get away with a cheaper resistor.

This explain why phase chopping is not suitable for unipolar motors : when the A winding is driven the chopper acts on the B winding. Clearly, this is no use at all for a variable reluctance motor and would be slow and inefficient for a bifilar wound permanent magnet motor.

The alternative is to tie the CONTROL input to ground so that the chopper acts on $\overline{INH1}$ and $\overline{INH2}$. Looking at the same example, A is high and B low. Q1 and Q4 are therefore conducting and current flows through Q1, the winding, Q4 and R_s, (figure 17).

Figure 15 : Phase Chopping. In this example the current X is interrupted by activating B, giving the recirculation path Y. The alternative, de-activating A, would give the recirculation path Z, increasing dissipation in R_s.

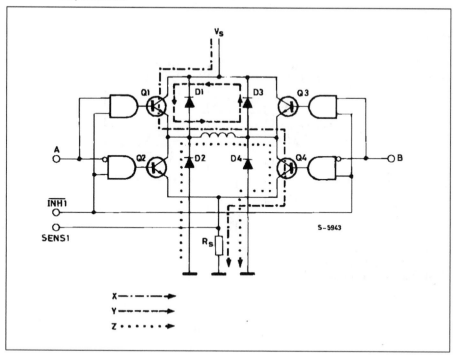

SGS-THOMSON
MICROELECTRONICS

Figure 16 : Phase Chopping Waveforms. The example shows AB winding energized with A positive with respect to B. Control is high.

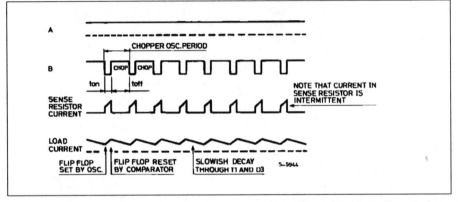

Figure 17 : Inhibit Chopping. The drive current (Q1, winding, Q4) in this case is interrupted by activating INH1. The decay path through D2 and D3 is faster than the path Y of Figure 15.

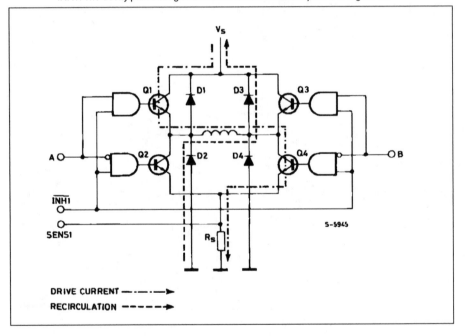

APPLICATION NOTE

In this case when the voltage accross R_S reaches V_{REF} the chopper flip flop is reset and $\overline{INH1}$ activated (brought low). $\overline{INH1}$, remember, turns off all four transistors therefore the current recirculates from ground, through D2, the winding and D3 to V_S. Discharged across the supply, which can be up to 46V, the current decays very rapidly (figure 18).

The usefulness of this second faster decay option is fairly obvious ; it allows fast operation with bipolar motors and it is the only choice for unipolar motors. But why do we offer the slower alternative, phase chopping ?

The answer is that we might be obliged to use a low chopper rate with a motor that does not store much energy in the windings. If the decay is very fast the average motor current may be too low to give an useful torque. Low chopper rates may, for example, be imposed if there is a larger motor in the same system. To avoid switching noise on the ground plane all drivers should be synchronized and the chopper rate is therefore determined by the largest motor in the system.

Multiple L297s are synchronized easily using the SYNC pin. This pin is the squarewave output of the on-chip oscillator and the clock input for the choppers. The first L297 is fitted with the oscillator components and outputs a sqarewave signal on this pin (figure 19). Subsequent L297s do not need the oscillator components and use SYNC as a clock input. An external clock may also be injected at this terminal if an L297 must be synchronized to other system components.

Figure 18 : Inhibit Chopper Waveforms. Winding AB is energized and CONTROL is low.

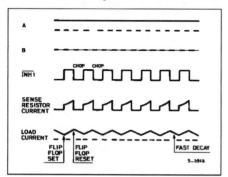

Figure 19 : The Chopper oscillator of multiple L297s are synchronized by connecting the SYNC Inputs together.

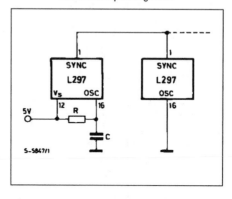

THE L297A

The L297A is a special version of the L297 developed originally for head positioning in floppy disk drives. It can, however, be used in other applications.

Compared to the standard L297 the difference are the addition of a pulse doubler on the step clock input and the availability of the output of the direction flip flop (block diagram, figure 20). To add these functions while keeping the low-cost 20-pin package the CONTROL and SYNC pins are not available on this version (they are note needed anyway). The chopper acts on the ABCD phase lines.

The pulse doubler generates a ghost pulse internally for each input clock pulse. Consequently the translator moves two steps for each input pulse. An external RC network sets the delay time between the input pulse and ghost pulse and should be chosen so that the ghost pulses fall roughly halfway between input pulses, allowing time for the motor to step.

This feature is used to improve positioning accuracy. Since the angular position error of a stepper motor is noncumulative (it cancels out to zero every four steps in a four step sequence motor) accuracy is improved by stepping two of four steps at a time.

SGS-THOMSON
MICROELECTRONICS

Figure 20 : The L297A, includes a clock pulse doubler and provides an output from the direction flip flop (DIR – MEM).

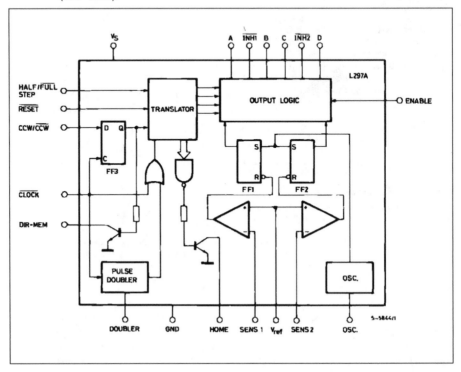

APPLICATION HINTS

Bipolar motors can be driven with an L297, an L298N or L293E bridge driver and very few external components (figure 21). Together these two chips form a complete microprocessor-to-stepper motor interface. With an L298N this configuration drives motors with winding currents up to 2A ; for motors up to 1A per winding and L293E is used. If the PWM choppers are not required an L293 could also be used (it doesn't have the external emitter connections for sensing resistors) but the L297 is underutilized. If very high powers are required the bridge driver is replaced by an equivalent circuit made with discrete transistors. For currents to 3.5A two L298N's with paralleled outputs may be used.

For unipolar motors the best choice is a quad darlington array. The L702B can be used if the choppers are not required but an ULN2075B is preferred.

This quad darlington has external emitter connections which are connected to sensing resistors (figure 22). Since the chopper acts on the inhibit lines, four AND gates must be added in this application.

Also shown in the schematic are the protection diodes.

In all applications where the choppers are not used it is important to remember that the sense inputs must be grounded and V_{REF} connected either to V_S or any potential between V_S and ground.

The chopper oscillator frequency is determined by the RC network on pin 16. The frequency is roughly $1/0.7$ RC and R must be more than 10 KΩ. When the L297A's pulse doubler is used, the delay time is determined by the network R_d C_d and is approximately 0.75 R_d C_d . R_d should be in the range 3 kΩ to 100 kΩ (figure 23).

 **SGS-THOMSON**
MICROELECTRONICS

APPLICATION NOTE

Figure 21 : This typical application shows an L297 and L298N driving a Bipolar Stepper Motor with phase currents up to 2A.

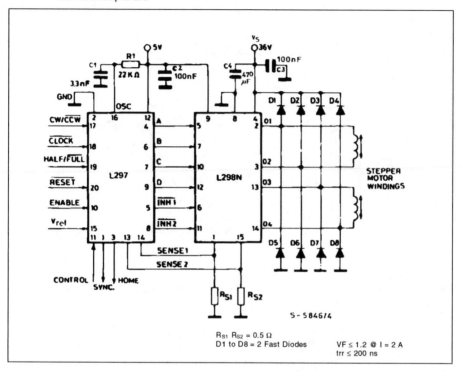

R_{S1} R_{S2} = 0.5 Ω
D1 to D8 = 2 Fast Diodes

VF ≤ 1.2 @ I = 2 A
trr ≤ 200 ns

SGS-THOMSON
MICROELECTRONICS

Figure 22 : For Unipolar Motors a Quad Darlington Array is coupled to the L297. Inhibit chopping is used so the four AND gates must be added.

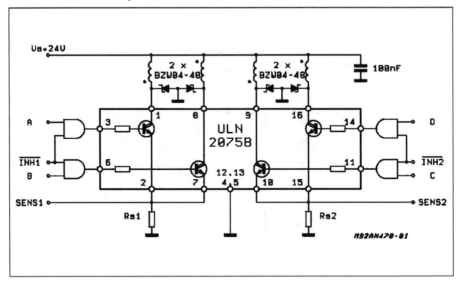

Figure 23 : The Clock pulse doubler inserts a ghost pulse τ_0 seconds after the Input clock pulse. R_d C_d is closen to give a delay of approximately half the Input clock period.

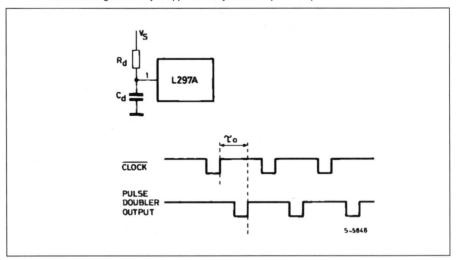

SGS-THOMSON
MICROELECTRONICS

APPLICATION NOTE

PIN FUNCTIONS - L297

N°	NAME	FUNCTION
1	SYNC	Output of the on-chip chopper oscillator. The SYNC connections The SYNC connections of all L297s to be synchronized are connected together and the oscillator components are omitted on all but one. If an external clock source is used it is injected at this terminal.
2	GND	Ground connection.
3	HOME	Open collector output that indicates when the L297 is in its initial state (ABCD = 0101). The transistor is open when this signal is active.
4	A	Motor phase A drive signal for power stage.
5	$\overline{INH1}$	Active low inhibit control for driver stage of A and B phases. When a bipolar bridge is used this signal can be used to ensure fast decay of load current when a winding is de-energized. Also used by chopper to regulate load current if CONTROL input is low.
6	B	Motor phase B drive signal for power stage.
7	C	Motor phase C drive signal for power stage.
8	$\overline{INH2}$	Active low inhibit control for drive stages of C and D phases. Same functions as INH1.
9	D	Motor phase D drive signal for power stage.
10	ENABLE	Chip enable input. When low (inactive) INH1, INH2, A, B, C and D are brought low.
11	CONTROL	Control input that defines action of chopper. When low chopper acts on INH1 and INH2; when high chopper acts on phase lines ABCD.
12	V_s	5V supply input.
13	$SENS_2$	Input for load current sense voltage from power stages of phases C and D.
14	$SENS_1$	Input for load current sense voltage from power stages of phases A and B.
15	V_{ref}	Reference voltage for chopper circuit. A voltage applied to this pin determines the peak load current.
16	OSC	An RC network (R to V_{CC}, C to ground) connected to this terminal determines the chopper rate. This terminal is connected to ground on all but one device in synchronized multi - L297 configurations. $f \cong 1/0.69\ RC$
17	CW/CCW	Clockwise/counterclockwise direction control input. Physical direction of motor rotation also depends on connection of windings. Synchronized internally therefore direction can be changed at any time.
18	$\overline{CLOCK}$	Step clock. An active low pulse on this input advances the motor one increment. The step occurs on the rising edge of this signal.
19	HALF/$\overline{FULL}$	Half/full step select input. When high selects half step operation, when low selects full step operation. One-phase-on full step mode is obtained by selecting FULL when the L297's translator is at an even-numbered state. Two-phase-on full step mode is set by selecting FULL when the translator is at an odd numbered position. (The home position is designate state 1).
20	$\overline{RESET}$	Reset input. An active low pulse on this input restores the translator to the home position (state 1, ABCD = 0101).

PIN FUNCTIONS - L297A (Pin function of the L297A are identical to those of the,L297 except for pins 1 and 11)

N°	NAME	FUNCTION
1	DOUBLER	An RC network connected to this pin determines the delay between an input clock pulse and the corresponding ghost pulse.
11	DIR-MEM	Direction Memory. Inverted output of the direction flip flop. Open collector output.

SGS-THOMSON
MICROELECTRONICS

Figure 24 : Pin connections.

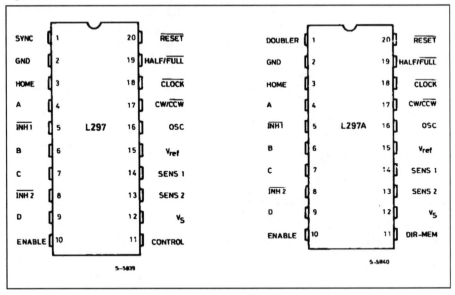

APPLICATION NOTE

SGS-THOMSON
MICROELECTRONICS

APPLICATION NOTE

STEPPER MOTOR DRIVER CONSIDERATIONS
COMMON PROBLEMS & SOLUTIONS

by Thomas L. Hopkins

This note explains how to avoid same of the more common pitfalls in motor drive design. It is based on the author's experience in responding to enquiries from the field.

INTRODUCTION

Over the years while working with stepper motor users, many of the same questions keep occurring from novice as well as experienced users of stepper motors. This application note is intended as a collection of answers to commonly asked questions about stepper motors and driver design. In addition the reference list contains a number of other application notes, books and articles that a designer may find useful in applying stepper motors.

Throughout the course of this discussion the reader will find references to the L6201, L6202 and L6203. Since these devices are the same die and differ only in package, any reference to one of the devices should be considered to mean any of the three devices.

Motor Selection (Unipolar vs Bipolar)

Stepper motors in common use can be divided into general classes, Unipolar driven motors and

Bipolar driven motors. In the past unipolar motors were common and preferred for their simple drive configurations. However, with the advent of cost effective integrated drivers, bipolar motors are now more common. These bipolar motors typically produce a higher torque in a given form factor [1].

Drive Topology Selection

Depending on the torque and speed required from a stepper motor there are several motor drive topologies available [5, chapter3]. At low speeds a simple direct voltage drive, giving the motor just sufficient voltage so that the internal resistance of the motor limits the current to the allowed value as shown in Figure 1A, may be sufficient. However at higher rotational speeds there is a significant fall off of torque since the winding inductance limits the rate of change of the current and the current can no longer reach it's full value in each step, as shown in Figure 2.

Figure 1: Simple direct voltage unipolar motors drive.

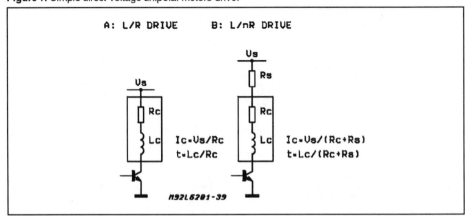

APPLICATION NOTE

Figure 2: Direct voltage drive.
A - low speed;
B - too high speed generates fall of torque.

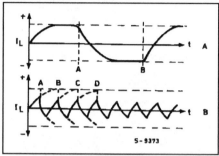

One solution is to use what is commonly referred to as an L/nR drive (Fig. 1B). In this topology a higher voltage is used and the current limit is set by an external resistor in series with the motor winding such that the sum of the external resistance and the internal winding resistance limits the current to the allowed value. This drive technique increases the current slew rate and typically provides better torque at high rotational speed. However there is a significant penalty paid in additional dissipation in the external resistances.

To avoid the additional dissipation a chopping controlled current drive may be employed, as shown in Figure 3. In this technique the current through the motor is sensed and controlled by a chopping control circuit so that it is maintained within the rated level. Devices like the L297, L6506 and PBL3717A implement this type of control. This technique improves the current rise time in the motor and improves the torque at high speeds while maintaining a high efficiency in the drive [2]. Figure 4 shows a comparison between the winding current wave forms for the same motor driven in these three techniques.

Figure 3: Chopper drive provides better performance.

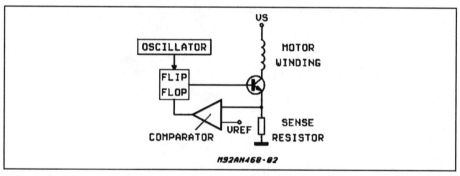

Figure 4: Motor current using L/R, L/5R and chopper constant current drive.

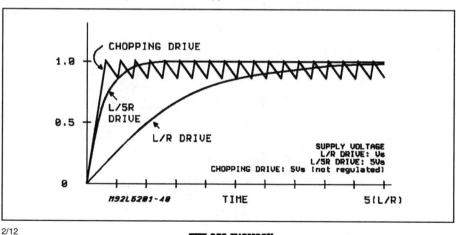

SGS-THOMSON
MICROELECTRONICS

In general the best performance, in terms of torque, is achieved using the chopping current control technique [2]. This technique also allows easy implementation of multiple current level drive techniques to improve the motor performance. [1]

Driving a Unipolar Motor with the L298N or L6202

Although it is not the optimal solution, design constraints sometimes limit the motor selection. In the case where the designer is looking for a highly integrated drive stage with improved performance over previous designs but is constrained to drive a unipolar wound (6 leaded) motor it is possible to drive the motor with H-Bridge drivers like the L298N or L6202. To drive such a motor the center tap of the motor should be left unconnected and the two ends of the common windings are connected to the bridge outputs, as shown in Figure 5. In this configuration the user should notice a marked improvement in torque for the same coil current, or put another way, the same torque output will be achieved with a lower coil current.

A solution where the L298N or L6202 is used to drive a unipolar motor while keeping the center connection of each coil connected to the supply will not work. First, the protection diodes needed from collector to emitter (drain to source) of the

bridge transistors will be forward biased by the transformer action of the motor windings, providing an effective short circuit across the supply. Secondly the L298N, even though it has split supply voltages, may not be used without a high voltage supply on the chip since a portion of the drive current for the output bridge is derived from this supply.

Selecting Enable or Phase chopping

When implementing chopping control of the current in a stepper motor, there are several ways in which the current control can be implemented. A bridge output, like the L6202 or L298N, may be driven in enable chopping, one phase chopping or two phase chopping, as shown in Figure 6. The L297 implements enable chopping or one phase chopping, selected by the control input. The L6506 implements one phase chopping, with the recirculation path around the lower half of the bridge, if the four outputs are connected to the 4 inputs of the bridge or enable chopping if the odd numbered outputs are connected to the enable inputs of the bridge. Selecting the correct chopping mode is an important consideration that affects the stability of the system as well as the dissipation. Table 1 shows a relative comparison of the different chopping modes, for a fixed chopping frequency, motor current and motor inductance.

Figure 5: Driving a unipolar wound motor with a bipolar drive

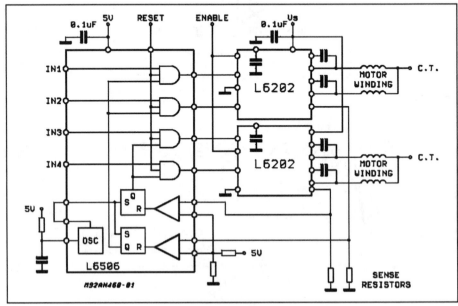

SGS-THOMSON MICROELECTRONICS

APPLICATION NOTE

Table 1: Comparative advantages of chopping modes

Chopping Mode	Ripple Current	Motor Dissipation	Bridge Dissipation *	Minimum Current
ENABLE	HIGH	HIGH	HIGH	LOWER
ONE PHASE	LOW	LOW	LOWEST	LOW
TWO PHASE	HIGH	LOW	LOW	Ipp/2

(*) As related to L298N, L6203 or L6202.

Figure 6a: Two Phase Chopping.

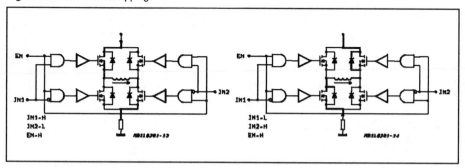

Figure 6b: One Phase Chopping.

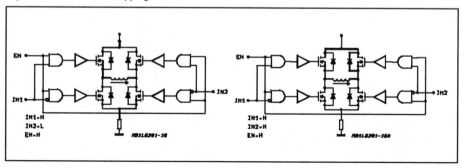

Figure 6c: Enable Chopping.

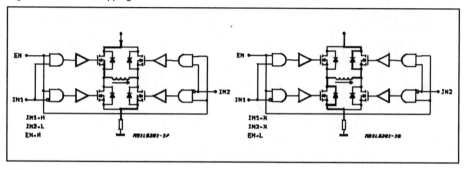

SGS-THOMSON
MICROELECTRONICS

RIPPLE CURRENT

Since the rate of current change is related directly to the voltage applied across the coil by the equation:

$$V = L \frac{di}{dt}$$

the ripple current will be determined primarily by the chopping frequency and the voltage across the coil. When the coil is driven on, the voltage across the coil is fixed by the power supply minus the saturation voltages of the driver. On the other hand the voltage across the coil during the recirculation time depends on the chopping mode chosen.

When enable chopping or two phase chopping is selected, the voltage across the coil during recirculation is the supply voltage plus either the V_F of the diodes or the RI voltage of the DMOS devices (when using the L6202 in two phase chopping). In this case the slope of the current rise and decay are nearly the same and the ripple current can be large.

When one phase chopping is used, the voltage across the coil during recirculation is V_{on} (V_{sat} for Bipolar devices or $I \cdot R_{DSon}$ for DMOS) of the transistor that remains on plus V_F of one diode plus the voltage drop across the sense resistor, if it is in the recirculation path. In this case the current decays much slower than it rises and the ripple current is much smaller than in the previous case. The effect will be much more noticeable at higher supply voltages.

MOTOR LOSSES

The losses in the motor include the resistive losses (I^2R) in the motor winding and parasitic losses like eddie current losses. The latter group of parasitic losses generally increases with increased ripple currents and frequency. Chopping techniques that have a high ripple current will have higher losses in the motor. Enable or two phase chopping will cause higher losses in the motor with the effect of raising motor temperature. Generally lower motor losses are achieved using phase chopping.

POWER DISSIPATION IN THE BRIDGE IC.

In the L298N, the internal drive circuitry provides active turn off for the output devices when the outputs are switched in response to the 4 phase inputs. However when the outputs are switched off in response to the enable inputs all base drive is removed from output devices but no active element is present to remove the stored charge in the base. When enable chopping is used the fall time of the current in the power devices will be longer and the device will have higher switching losses than if phase chopping is used.

In the L6202 and L6203, the internal gate drive circuit works the same in response to either the input or the enable so the switching losses are the same using enable or two phase chopping, but would be lower using one phase chopping. However, the losses due to the voltage drops across the device are not the same. During enable chopping all four of the output DMOS devices are turned off and the current recirculates through the body to drain diodes of the DMOS output transistors. When phase chopping the DMOS devices in the recirculation path are driven on and conduct current in the reverse direction. Since the voltage drop across the DMOS device is less than the forward voltage drop of the diode for currents less than 2A, the DMOS take a significant amount of the current and the power dissipation is much lower using phase chopping than enable chopping, as can be seen in the power dissipation graphs in the data sheet.

With these two devices, phase chopping will always provide lower dissipation in the device. For discrete bridges the switching loss and saturation losses should be evaluated to determine which is lower.

MINIMUM CURRENT

The minimum current that can be regulated is important when implementing microstepping, when implementing multilevel current controls, or anytime when attempting to regulate a current that is very small compared to the peak current that would flow if the motor were connected directly to the supply voltage used.

With enable chopping or one phase chopping the only problem is loss of regulation for currents below a minimum value. Figure 7 shows a typical response curve for output current as a function of the set reference. This minimum value is set by the motor characteristics, primarily the motor resistance, the supply voltage and the minimum duty cycle achievable by the control circuit. The minimum current that can be supplied is the current that flows through the winding when driven by the minimum duty cycle. Below this value current regulation is not possible. With enable chopping the current through the coil in response to the minimum duty cycle can return completely to zero during each cycle, as shown in figure 8. When using one phase chopping the current may or may not return completely to zero and there may be some residual DC component.

When using a constant frequency control like the L297 or L6506, the minimum duty cycle is basically the duty cycle of the oscillator (sync) since the set dominance of the flip-flop maintains the output on during the time the sync is active. In constant off time regulators, like the PBL3717A, the minimum output time is set by the propagation delay through the circuit and it's ratio to the selected off time.

SGS-THOMSON
MICROELECTRONICS

APPLICATION NOTE

Figure 7: The transfer function of peak detect current control is nonlinear for low current values.

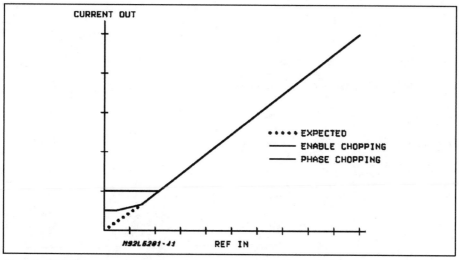

Figure 8: A Minimum current flows through the motor when the driver outputs the minimum duty cycle that is achievable.

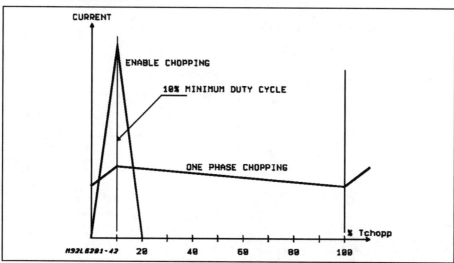

For two phase chopping the situation is quite different. Although none of the available control chips implement this mode it is discussed here since it is easy to generate currents that can be catastrophic if two phase chopping is used with peak detecting control techniques. When the peak current is less than 1/2 of the ripple (I_{pp}) current two phase chopping can be especially dangerous. In this case the reverse drive ability of the two phase chopping technique can cause the current in the motor winding to reverse and the control circuit to lose control. Figure 9 shows the current wave form in this case. When the current reaches the peak set by the reference both sides of the bridge are switched and the current decays until it reaches zero. Since the power transistors

SGS-THOMSON
MICROELECTRONICS

Figure 9: Two phase chopping can loose control of the winding current..

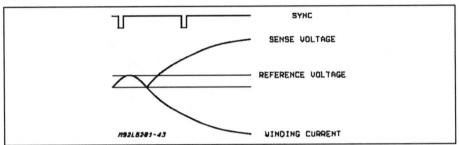

are now on, the current will begin to increase in a negative direction. When the oscillator again sets the flip-flop the inputs will then switch again and the current will begin to become more positive. However, the effect of a single sense resistor used with a bridge is to rectify current and the comparator sees only the magnitude and not the sign of the current. If the absolute value of the current in the negative direction is above the set value the comparator will be fooled and reset the flip-flop. The current will continue to become more negative and will not be controlled by the regulation circuit.

For this reason two phase chopping is not recommended with bridge circuits like the L298N or L6203 and is not implemented in any of the currently available driver IC's. The problem can be avoided by more complex current sense techniques that do not rectify the current feedback.

Chopper Stability and Audio Noise.

One problem commonly encountered when using chopping current control is audio noise from the motor which is typically a high pitch squeal. In constant frequency PWM circuits this occurrence is usually traced to a stability problem in the current control circuit where the effective chopping frequency has shifted to a sub-harmonic of the desired frequency set by the oscillator. In constant off time circuits the off time is shifted to a multiple of the off time set by the monostable. There are two common causes for this occurrence.

The first cause is related to the electrical noise and current spikes in the application that can fool the current control circuit. In peak detect PWM circuits, like the L297 and L6506, the motor current is sensed by monitoring the voltage across the sense resistor connected to ground. When the oscillator sets the internal flip flop causing the bridge output to turn on, there is typically a voltage spike developed across this resistor. This spike is caused by noise in the system plus the reverse recovery current of the recirculating diode that flows through the sense resistor, as shown in

Figure 10. If the magnitude of this spike is high enough to exceed the reference voltage, the comparator can be fooled into resetting the flip-flop prematurely as shown in Figure 11. When this occurs the output is turned off and the current continues to decay. The result is that the fundamental frequency of the current wave form delivered to the motor is reduced to a sub-harmonic of the oscillator frequency, which is usually in the audio range. In practice it is not uncommon to encounter instances where the period of the current wave form is two, three or even four times the period of the oscillator. This problem is more pronounced in breadboard implementations where the ground is not well laid out and ground noise contributes makes the spike larger.

When using the L6506 and L298N, the magnitude of the spike should be, in theory, smaller since the diode reverse recovery current flows to ground and not through the sense resistor. How-

Figure 10: Reverse recovery current of the recirculation diode flows through the sense resistor causing a spike on the sense resistor.

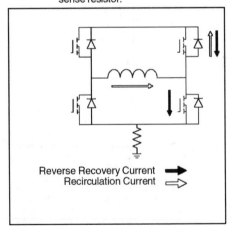

Reverse Recovery Current ➡
Recirculation Current ⇨

70

APPLICATION NOTE

Figure 11: Spikes on the sense resistor caused by reverse recovery currents and noise can trick the current sensing comparator.

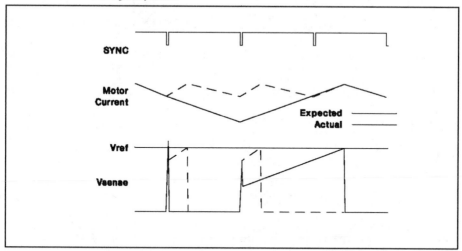

ever, in applications using monolithic bridge drivers, like the L298N, internal parasitic structures often produce recovery current spikes similar in nature to the diode reverse recovery current and these may flow through the emitter lead of the device and hence the sense resistor. When using DMOS drivers, like the L6202, the reverse recovery current always flows through the sense resistor since the internal diode in parallel with the lower transistor is connected to the source of the DMOS device and not to ground.

In constant off time FM control circuits, like the

CALCULATING POWER DISSIPATION IN BRIDGE DRIVER IC'S

The power dissipated in a monolithic driver IC like the L298N or L6202 is the sum of three elements: 1) the quiescent dissipation, 2) the saturation losses and 3) the switching losses.

The quiescent dissipation is basically the dissipation of the bias circuitry in the device and can be calculated as $Vs \cdot Is$ where Vs is the power supply voltage and Is is the bias current or quiescent current from the supply. When a device has two supply voltages, like the L298N, the dissipation for each must be calcualted then added to get the total quiescent dissipation. Generally the quiescent current for most monolithic IC's is constant over a vide range of input voltages and the maximum value given on the data sheet can be used for most supply voltages within the allowable range.

The saturation loss is basically the sum of the voltage drops times the current in each of the output transistors. For Bipolar devices, L298N, this is $Vsat \cdot I$. For DMOS power devices this is $I^2 \cdot R_{DSon}$.

The third main component of dissipation is the switching loss associated with the output devices. In general the switching loss can be calculated as:

$$Vsupply \cdot Iload \cdot tcross \cdot fswitch$$

To calculate the total power dissipation these three compnents are each calculated, multipled by their respective duty cycle then added togther. Obviously the duty cycle for the quiescent current is equal to 100%.

![SGS-THOMSON MICROELECTRONICS]

PBL3717A, the noise spike fools the comparator and retriggers the monostable effectively multiplying the set off time by some integer value.

Two easy solutions to this problem are possible. The first is to put a simple RC low pass filter between the sense resistor and the sense input of the comparator. The filter attenuates the spike so it is not detected by the comparator. This obviously requires the addition of 4 additional components for a typical stepper motor. The second solution is to use the inherent set dominance of the internal flip-flop in the L297 or L6506 [1][3] to mask out the spike. To do this the width of the oscillator sync pulse is set to be longer than the sum of the propagation delay (typically 2 to 3μs for the L298N) plus the duration of the spike (usually in the range of 100ns for acceptable fast recovery diodes), as shown in figure 12. When this pulse is applied to the flip-flop set input, any signal applied to the reset input by the comparator is ignored. After the set input has been removed the comparator can properly reset the flip-flop at the correct point.

The corresponding solution in frequency modulated circuits, is to fix a blanking time during which the monostable may not be retriggered.

The best way to evaluate the stability of the chopping circuit is to stop the motor movement (hold the clock of the L297 low or hold the four inputs constant with the L6506) and look at the current wave forms without any effects of the phase changes. This evaluation should be done for each level of current that will be regulated. A DC current probe, like the Tektronix AM503 system, provides the most accurate representation of the motor current. If the circuit is operating stability, the current wave form will be synchronized to the sync signal of the control circuit. Since the spikes discussed previously are extremely short, in the range of 50 to 150 ns, a high frequency scope with a bandwidth of at least 200 MHz is required to evaluate the circuit. The sync signal to the L297 or L6506 provides the best trigger for the scope.

The other issue that affects the stability of the constant frequency PWM circuits is the chopping mode selected. With the L297 the chopping signal may be applied to either the enable inputs or the four phase inputs. When chopping is done using the enable inputs the recirculation path for the current is from ground through the lower recirculation diode, the load, the upper recirculation diode and back to the supply, as shown in Figure 6c. This same recirculation path is achieved using two phase chopping, although this may not be im-

Figure 12: The set-dominant latch in the L297 may be used to mask spikes on the sense resistor that occur at switching.

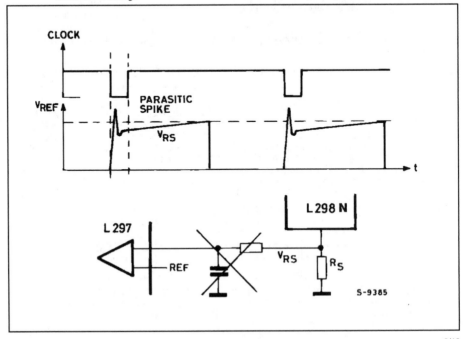

APPLICATION NOTE

plemented directly using the L297 or L6506. In this mode, ignoring back EMF, the voltage across the coil during the on time (t_1) when current is increasing and the recirculation time (t_2), are:

$$V_1 = V_s - 2 V_{sat} - V_{Rsense}$$

and

$$V_2 = V_{ss} + 2 V_F$$

The rate of current change is given by (ignoring the series resistance):

$$V = L \frac{di}{dt}$$

Since the voltage across the coil (V_2) during the recirculation time is more than the voltage (V_1) across the coil during the on time the duty cycle will, by definition, be greater than 50% because t_1 must be greater than t_2. When the back EMF of the motor is considered the duty cycle becomes even greater since the back EMF opposes the increase of current during the on time and aides the decay of current.

In this condition the control circuit may be content to operate stability at one half of the oscillator frequency, as shown in Figure 13. As in normal operation, the output is turned off when the current reaches the desired peak value and decays until the oscillator sets the flip-flop and the current again starts to increase. However since t_1 is longer than t_2 the current has not yet reached the peak value before the second oscillator pulse occurs. The second oscillator pulse then has no effect and current continues to increase until the set peak value is reached and the flip-flop is reset by

the comparator. The current control circuit is completely content to keep operating in this condition. In fact the circuit may operate on one of two stable conditions depending on the random time when the peak current is first reached relative to the oscillator period.

The easiest, and recommended, solution is to apply the chopping signal to only one of the phase inputs, as implemented with the L297, in the phase chopping mode, or the L6506.

Another solution that works, in some cases, is to fix a large minimum duty cycle, in the range of 30%, by applying an external clock signal to the sync input of the L297 or L6506. In this configuration the circuit must output at least the minimum duty cycle during each clock period. This forces the point where the peak current is detected to be later in each cycle and the chopping frequency to lock on the fundamental. The main disadvantage of this approach is that it sets a higher minimum current that can be controlled. The current in the motor also tends to overshoot during the first few chopping cycles since the actual peak current is not be sensed during the minimum duty cycle.

EFFECTS OF BACK EMF

As mentioned earlier, the back EMF in a stepper motor tends to increase the duty cycle of the chopping drive circuits since it opposes current increased and aids current decay. In extreme, cases where the power supply voltage is low compared to the peak back EMF of the motor, the duty cycle required when using the phase chopping may exceed 50% and the problem with the

Figure 13: When the output duty cycle exceeds 50% the chopping circuit may sinchronize of a sub-harmonic of the oscillator frequency.

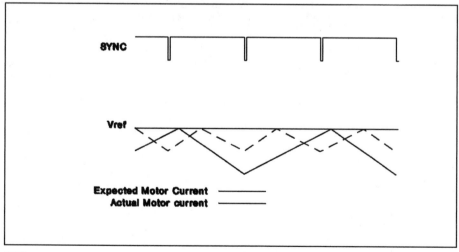

SGS-THOMSON
MICROELECTRONICS

stability of the operating frequency discussed above can occur. At this point the constant frequency chopping technique becomes impractical to implement and a chopping technique that uses constant off time frequency modulation like implemented in the PBL3717A, TEA3717, TEA3718, and L6219 is more useful.

Why Won't the motor move

Many first time users of chopping control drives first find that the motor does not move when the circuit is enabled. Simply put the motor is not generating sufficient torque to turn. Provided that the motor is capable of producing the required torque at the set speed, the problem usually lies in the current control circuit. As discussed in the previous section the current sensing circuit can be fooled. In extreme cases the noise is so large that the actual current through the motor is essentially zero and the motor is producing no torque. Another symptom of this is that the current being drawn from the power supply is very low.

Avoid Destroying the Driver

Many users have first ask why the device failed in the application. In almost every case the failure was caused by electrical overstress to the device, specifically voltages or currents that are outside of the device ratings. Whenever a driver fails, a careful evaluation of the operating conditions in the application is in order.

The most common failure encountered is the result of voltage transients generated by the inductance in the motor. A correctly designed application will keep the peak voltage on the power supply, across the collector to emitter of the output devices and, for monolithic drivers, from one output to the other within the maximum rating of the device. A proper design includes power supply filtering and clamp diodes and/or snubber networks on the output [6].

Selecting the correct clamp diodes for the application is essential. The proper diode is matched to the speed of the switching device and maintains a V_F that limits the peak voltage within the allowable limits. When the diodes are not integrated they must be provided externally. The diodes should have switching characteristics that are the same or better than the switching time of the output transistors. Usually diodes that have a reverse recovery time of less than 150 ns are sufficient when used with bipolar output devices like the L298N. The 1N4001 series of devices, for example, **is not** a good selection because it is a slow diode.

Although it occurs less frequently, excess current can also destroy the device. In most applications the excess current is the result of short circuits in the load. If the application is pron to have shorted loads the designer may consider implementing some external short circuit protection [7].

Shoot through current, the current that flows from supply to ground due to the simultaneous conduction of upper and lower transistors in the bridge output, is another concern. The design of the L298N, L293 and L6202 all include circuitry specifically to prevent this phenomena. The user should not mistake the reverse recovery current of the diodes or the parasitic structures in the output stage as shoot through current.

SELECTED REFERENCES

[1]Sax, Herbert., "Stepper Motor Driving" (AN235)

[2]"Constant Current Chopper Drive Ups Stepper-Motor Performance" (AN468)

[3]Hopkins, Thomas. "Unsing the L6506 for Current Control of Stepping Motors" (AN469)

[4]"The L297 Steper Motor Controller" (AN470)

[5]Leenouts, Albert. The Art and Practice of Step Motor Control. Ventura CA: Intertec Communications Inc. (805) 658-0933. 1987

[6]Hopkins, Thomas. "Controlling Voltage Transisnts in Full Bridge Drivers" (AN280)

[7]Scrocchi G. and Fusaroli G. "Short Circuit Protection on L6203". (AN279)

APPLICATION NOTE

3

Making the Printed Circuit Board

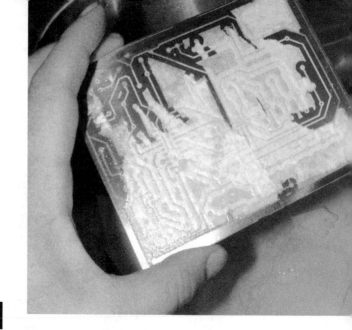

Tools and Material

In order to make the boards, you will need these items, as seen in
Figure 3.1:

- Copper-clad PCB material

- Glossy laser printer paper

- Permanent marker

- Iron

- Plastic or glass container

- Ferric chloride

- Thermometer

- Stapler

- Steel wool

- Kitchen sink

- Magnifying glass

Figure 3.1

The tools and material.

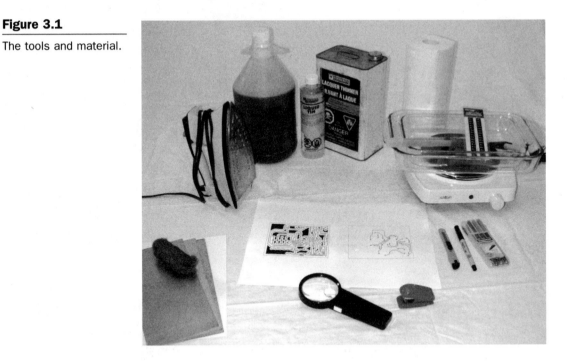

Artwork

Based on the schematics produced in the last chapter, I created the artwork for the printed circuit boards. This artwork must be reproduced on the surface of the copper in a manner that will enable it to act as a *resist* to the etching solution. In order to make the PCB, we must remove all the unneeded copper from the board, leaving us with the copper traces that take the place of wire. You can wire a circuit together using breadboards, but the time and effort are better spent making a PCB to suit. I did build this circuit first using a breadboard but only to make sure it was really going to work. There are options for the method you employ in transferring the art to the copper. The most popular is the photo-resist method. Electronics suppliers generally sell kits containing the chemistry and tools for exposing and developing presensitized boards.

If you plan on using photo resist, you can scan the artwork and print the transparencies needed. I won't expand on the photo-resist method because detailed instructions are included with the packaged material.

I make my boards using the toner-transfer method. It does have drawbacks, the most obvious being trace size. You can't make the traces too small or close together. If you do, they will probably short out because it's not easy to get the leftover paper off the board after ironing. As a result, the PCB tends to be a little bigger than one made with photo resist. This project doesn't require small components, so don't worry about the size of the boards.

I created a double-sided board for the driver layout (see **Figure 3.2**) but don't feel obliged to make a double-sided board if you don't want to, or can't secure the material. Use the topside art (see

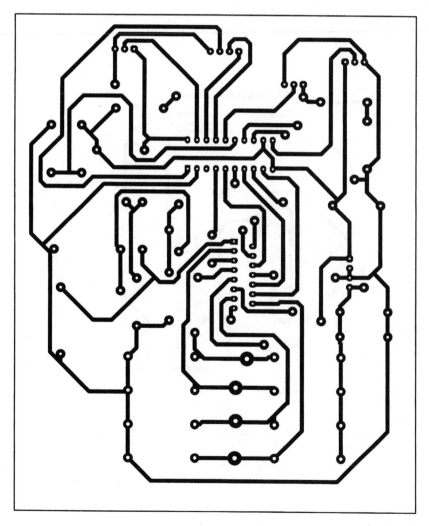

Figure 3.2

Bottom of driver board.

Figure 3.3) as a wiring diagram for jumpers if you make a single-sided board. The interface board is single sided (see **Figure 3.4**), with only one jumper on the topside.

Scan the artwork into an image-editing program, making sure you preserve the size exactly as published. Then print it with a laser printer. Make certain that the printer is set up to print at its highest resolution. You may also be able to print darker if your printer will let you make the adjustments. The idea is to print the art with lots of toner so when it gets transferred there are no gaps or holes exposing copper we want to keep. I use an HP 4 laser printer; it's

Figure 3.3

Top of driver board, this image has been mirrored already.

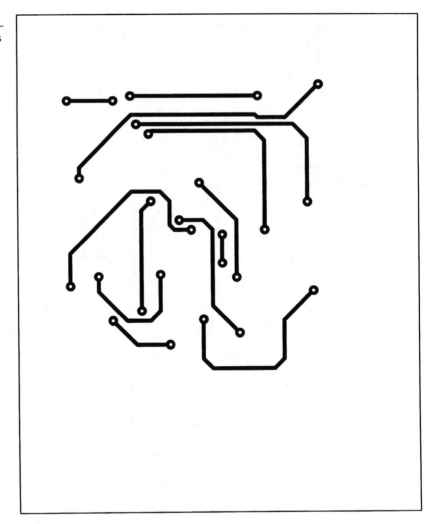

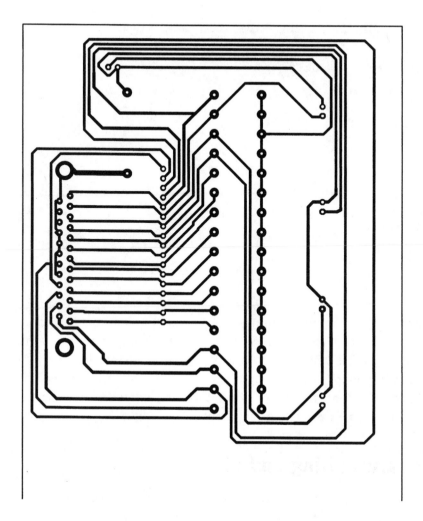

Figure 3.4

Bottom of interface board.

old and slow but still gets the job done. In **Figure 3.5** you can see the settings for my printer.

The next step is to print the board artwork on glossy laser printer paper. I'm using paper that I purchased from a local paper supplier that does most of its business with print shops. They didn't mind selling me a small quantity of glossy laser paper; in fact they were very helpful. I admit I wasted some money experimenting with paper I purchased from big-box retail stores. Don't even entertain the notion of using ink jet paper in a laser printer; it may result in some expensive repair bills. I did try glossy ink jet paper

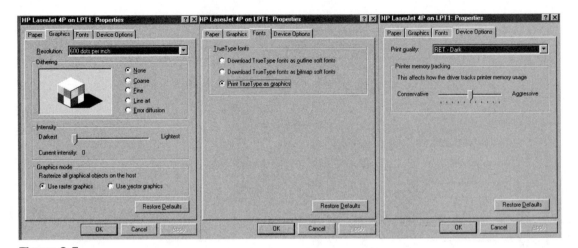

Figure 3.5

Printer driver setup.

printed from an ink jet printer, but I wasn't able to get the paper off the board. With heat, the surface of the ink jet paper fuses to the copper. If you have problems getting the paper to free itself from the board after ironing, then a little experimentation will be in order. You might be able to print the artwork with a photocopier as long as it will use glossy paper.

Board Cutting and Cleaning

After printing the artwork, I decide how big a piece of copper board I need, then cut it on my little table saw using a nonferrous blade (see **Figure 3.6**).

Because I have a tendency to purchase material considered surplus, I generally have a little extra work before I can use it. In this case the PCB material is oxidized so I will clean it up with some superfine steel wool (see **Figure 3.7**).

After the steel wool treatment, I wash the board with dish soap and hot water (as seen in **Figure 3.8**), making sure I keep my hands off of the copper.

If you aren't careful when handling the board, your cleaning will be to no avail because the oil from your fingers will inhibit the

Figure 3.6

Cutting PCB material.

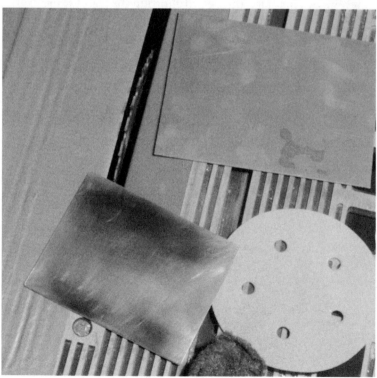

Figure 3.7

Shiny copper.

Figure 3.8

Washing the board.

laser toner from adhering to the copper and could act as a resist when etching begins. Dry the board thoroughly (see **Figure 3.9**).

Figure 3.9

Drying with paper towel.

Toner Transfer

After printing the artwork, you will need to align the two pieces of paper *ink side in* so that the toner makes contact with the copper on the board. Use a strong light source or hold the paper up to a window to accomplish this task. Or use a light box if you have one (see **Figures** 3.10 and 3.11).

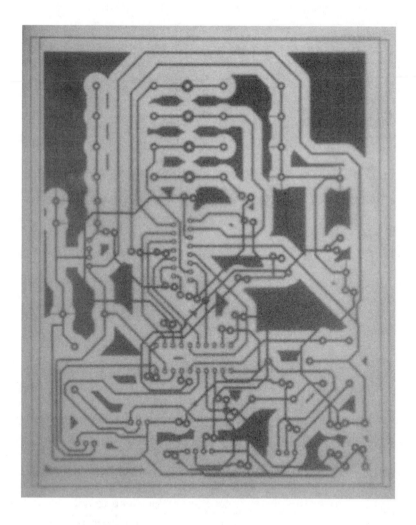

Figure 3.10

Unaligned art work.

Figure 3.11

Aligned artwork.

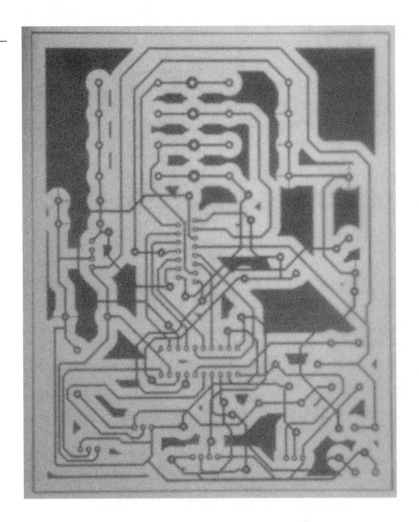

Now staple three sides of the sandwiched paper to maintain alignment (see **Figure 3.12**).

Place the paper sandwich that includes the clean piece of copper-clad board, making sure the toner is aligned within the copper board (see **Figure 3.13**).

Having heated the iron to its highest setting, place a clean piece of regular paper over the glossy laser paper. You need to use the regular paper so that the iron won't stick to the laser paper (see **Figure 3.14**).

Figure 3.12

Fixing position of art with staples.

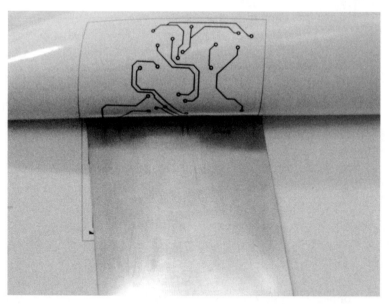

Figure 3.13

Inserting the board.

If you don't use something between the glossy paper and the iron, you will make a mess of both of them. I like to iron the first side a little in order to fix it in position. Then I flip over the whole sandwich and start to iron the second side using moderate pressure for a minute or longer, trying to make sure that I give equal attention to all areas of the board. Next, flip it all over and finish the first side (see **Figure 3.15**).

Figure 3.14

Using plain paper
barrier.

Figure 3.15

Ironing the board.

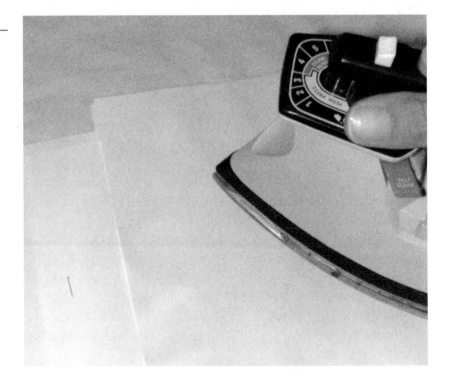

In order to set the toner, you must rinse or immerse the sandwich in cold water for a couple of minutes (see **Figure 3.16**).

Figure 3.16

Cold rinse.

Soak everything in warm water for five or ten minutes. After a good soaking, peel the paper from both sides of the board (see **Figures 3.17** and **3.18**).

If the residual paper doesn't rub off easily with a little thumb pressure, then resoak the board to loosen it up (see **Figure 3.19**).

After rubbing off the residual paper, thoroughly check the toner for missing sections. If you find any, touch up the areas with a permanent marker (see **Figures 3.20** and **3.21**).

Figure 3.17

Paper peeling.

Figure 3.18

Thumb rubbing.

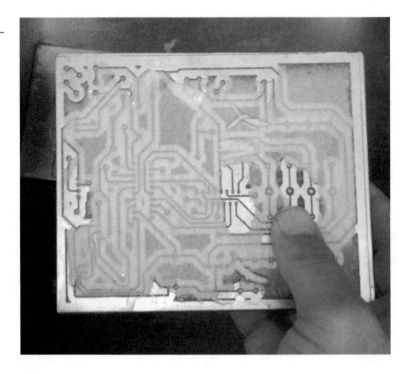

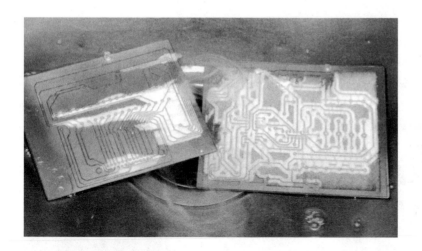

Figure 3.19

Re-soak.

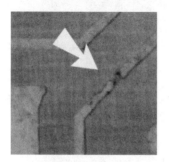

Figure 3.20

Missing toner.

Figures 3.21

Touchup.

Go over the areas a couple of times to make sure the copper is well protected. After looking for missing toner, check the holes in the pads and remove any residual paper by gently scratching it with a knife tip or a pin (see **Figures** 3.22 and 3.23).

Etching the copper out of the pad holes makes drilling a lot more pleasant. It's handy to use a magnifying glass for the inspection and touch up.

Figure 3.22

Paper in hole.

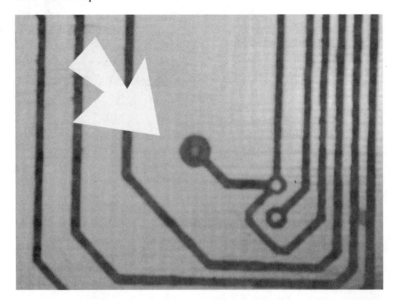

Figure 3.23

Scratch it out.

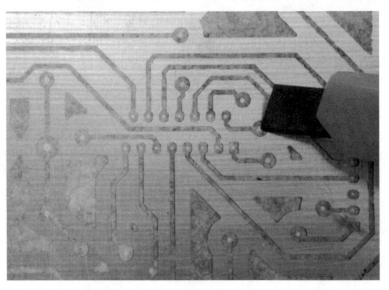

Etching

When you are happy with the toner resist, proceed to etch the boards. Ferric chloride will remove all the unprotected copper on the board. **WARNING:** It will also burn your skin, so make sure you use gloves, eye protection, and a respirator when working with this etching chemistry. Try to maintain a good working temperature with the setup (see **Figure 3.24**). Don't let the temperature of the working solution go higher than 40 degrees Celsius (104 degrees Fahrenheit). At temperatures higher than that, it will give off some nasty and dangerous fumes.

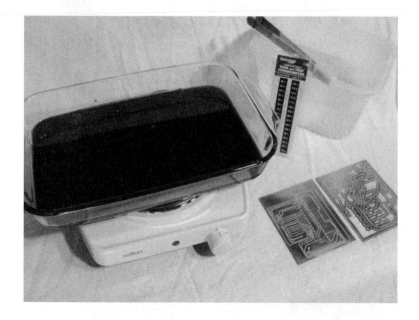

Figure 3.24

Etching setup.

I like to use a floating thermometer to keep an eye on the temperature, but any thermometer should work. Keep checking the board in the ferric chloride solution (see **Figures 3.25** and **3.26**).

The amount of time required to etch the board will vary with the temperature of the ferric chloride and how many times the etching solution has been used. When the board looks like **Figure 3.26**, it's done. Rinse the board in clean water to stop the etching (see **Figure 3.27**).

Figure 3.25

Not finished.

Figure 3.26

Finished.

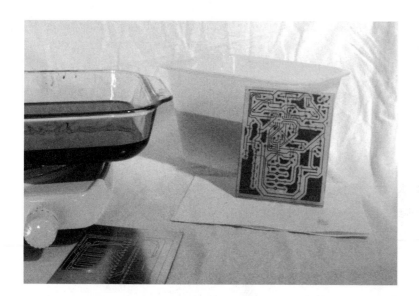

Figure 3.27

Rinse.

You may need to use a foam paintbrush to spot etch an area if the rest of the board is finished (see **Figure 3.28**).

Figure 3.28

Spot etching.

Now, toner with copper underneath is all that remains on the board. To clean off the toner, I use a rag with some lacquer thinner and possibly a little soaking in a shallow tray of the thinner (see **Figure 3.29**).

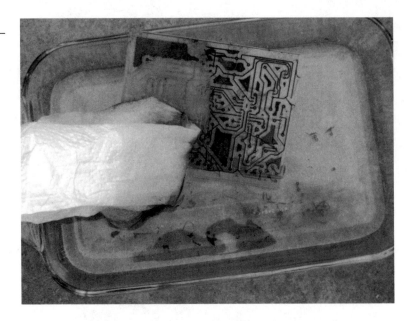

The next step isn't necessary, but I like the look of a board that has been tinned. If you want to *tin* the board, pick up some tinning solution and set the clean board in a shallow tray with enough solution to cover it. Wait until the copper has a uniform deposit and remove and wash the board. The added benefit to this application of tin is that soldering will be a little easier (see **Figure 3.30**).

These are the finished boards, ready for drilling (see **Figure 3.31**).

In this chapter you covered how to manufacture printed circuit boards for your CNC project, sources for materials, as well as what methods work best. Now you should have three driver boards and one interface board etched. Etching out of the way, the next chapter shows you how to drill the required holes in your boards and finish assembling them.

Figure 3.30

Tinning the boards.

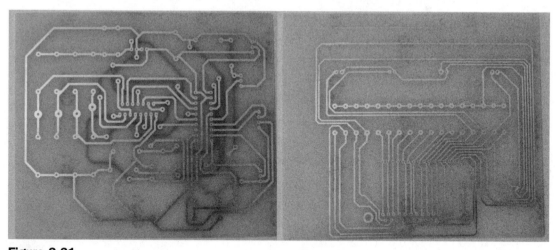

Figure 3.31

Etched and tinned.

4
Driver Assembly

In order to build the driver assembly for the CNC machine, you will need these items, as seen in **Figure 4.1**:

- Three etched driver boards

- One etched interface board

- Drill press

- Drill bits

- Components for each board

- Soldering iron

- Anti-static wrist band

- Pliers

- Cutter

- Wire stripper

- Desoldering tool

- Multimeter

- Solder

Figure 4.1

Tools and material required.

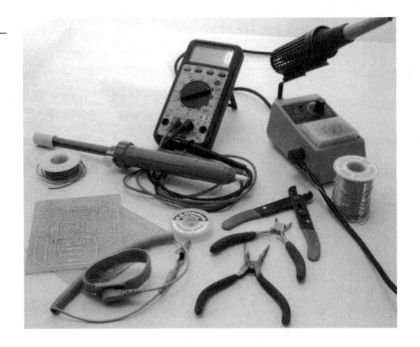

Drilling the holes for the components is the first step in the assembly process. This is also the most tedious and least-liked part of any electronics project, but a small drill press like the one in **Figure** 4.2 and some sharp drill bits can make a big difference.

Figure 4.2

A small drill press.

When the boards were etched, the hole locations in the pads were etched leaving the fiberglass exposed. You will find that the drill bit will center itself to the bare fiberglass, allowing you to drill a little faster and with more accuracy than if the hole locations had not been etched free of copper (see **Figure 4.3**).

Figure 4.3

Closeup of pad hole.

For the resistors, IC locations, and the through-holes I used a #60 drill bit. For the diode lead holes and the motor wire locations I used a 1/16-inch drill bit. On the interface board, use a 1/8-inch drill bit for the DB25 mounting holes (see **Figure 4.4**).

When drilling the holes, it's helpful to place a piece of scrap wood under the board so as not to damage any of the traces. See **Figure 4.5**.

Once you have drilled the component lead holes, turn on your soldering iron. If you have a variable-temperature soldering iron, find a temperature that will allow you to solder quickly but not be so hot as to lift the pads from the fiberglass (see **Figure 4.6**).

The method I use to solder is to hold the tip of the iron to a place on the pad where I can also make contact with the lead; then I apply solder to the pad and lead junction opposite the soldering iron tip (see **Figure 4.7**).

Figure 4.4

Drill bits required.

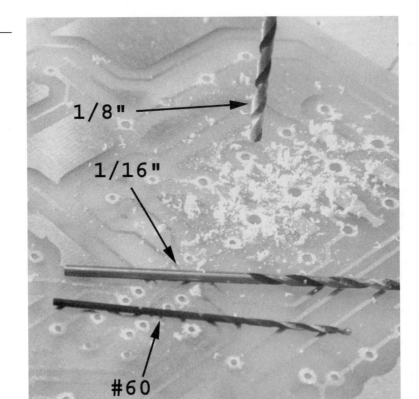

1/8"

1/16"

#60

Figure 4.5

Drilling holes in PCB.

Figure 4.6

Soldering iron.

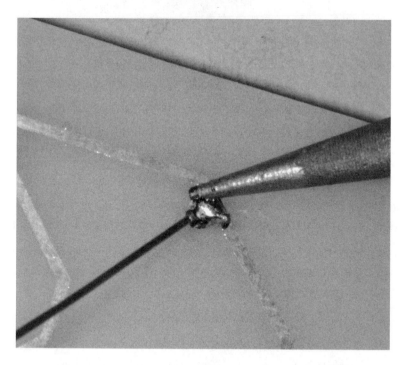

Figure 4.7

Soldering technique.

The solder will draw around the lead and pad to the source of heat. Before you touch any of the ICs make sure you have some kind of static protection in place. I use a wrist strap connected to ground at the back of my soldering iron (see **Figure 4.8**).

Figure 4.8

Wrist strap static
protection.

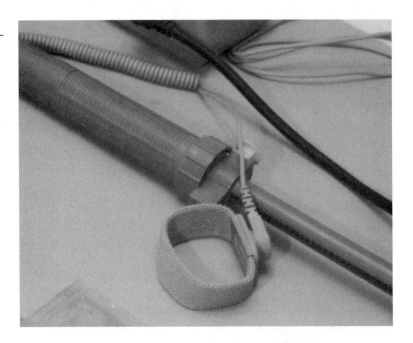

Static electricity will destroy the integrated circuits. Refer to **Figure 4.9** to determine the placement of the components on the driver board or **Figure 4.10** for the interface component placement.

Start by installing the jumpers on the topside of the board. You can use insulated wire (see **Figure 4.11**) to make the connections from one end of the top traces to the other end or you can use a piece of solid wire or any material that can act as a conductor from one side of the through-hole to the other. On one board, I used pieces of a paper clip cut to extend a little past each side of the board (see **Figure 4.12**).

This method of mimicking a plated through-hole requires the conducting material to be soldered on both sides of the board and can be a little tricky if the material doesn't fit snugly in the hole. If the material is a little loose, it has a tendency to slide through and may not actually create a connection between top and bottom. You can as easily use wire bent on both sides of the through-hole. If you use lengths of insulated wire to follow the topside traces, you only need to solder at the bottom of the board because the top traces have become jumper guides as opposed to being the jumpers. After installing the jumpers, check with a multimeter to ensure that continuity exists between the connections (see **Figure 4.13**).

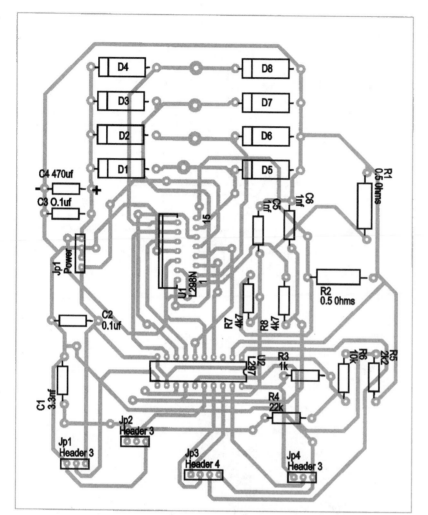

Figure 4.9

Driver board part placement.

Now start installing the components that will be lowest on the board—resistors and diodes. The resistors don't have to be installed with polarity in mind, but the diodes will need to be placed with the band on their body as in **Figure 4.9**. When you are bending the leads of the components, hold the lead close to the body of the component with pliers and bend the lead (see **Figure 4.14**). This method avoids undue stress on the body of the component.

After inserting the leads into the appropriate holes on the board, bend them over and trim them to enough length to hold the component in place. The diodes' leads should be trimmed as short as

Figure 4.10

Interface board part placement.

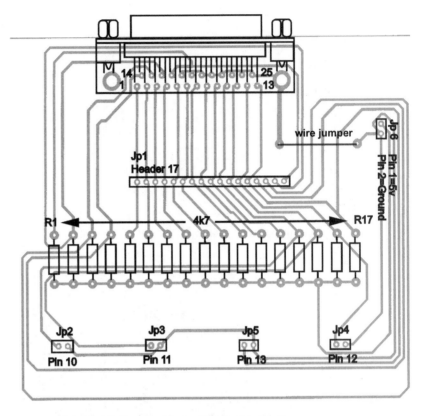

Figure 4.11

Using insulated wire.

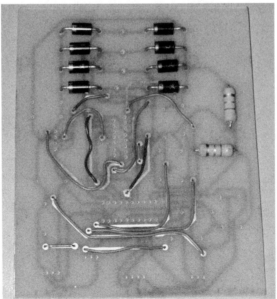

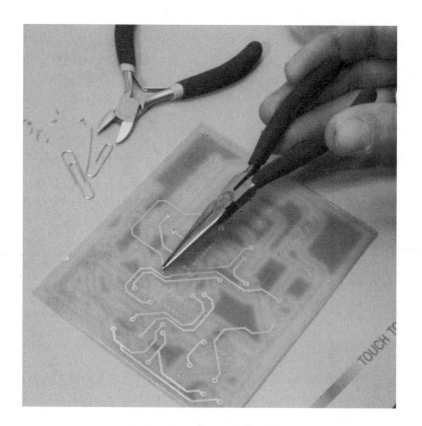

Figure 4.12

Using pieces of paper clip.

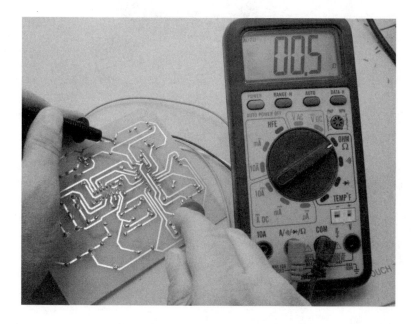

Figure 4.13

Checking continuity with a multimeter.

Figure 4.14

Bending a component
lead.

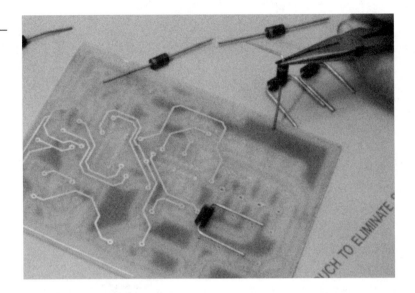

possible. Their size makes them more difficult to solder, because it takes more time to heat them up to the point at which they will accept solder (see **Figure 4.15**).

After installing the short components, I installed the headers, cut the header material to the necessary size, and soldered (see **Figure 4.16**).

Figure 4.15

Bending and trimming
leads.

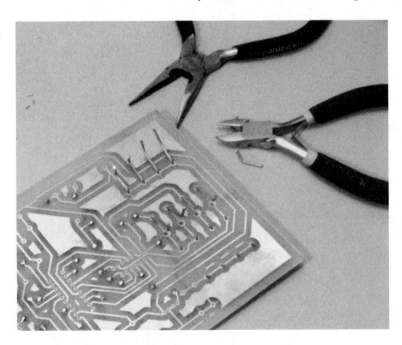

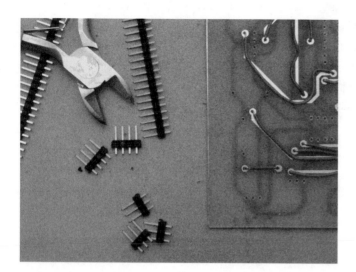

Figure 4.16

Cutting header material.

To install the IC socket for the L297, bend a couple of the leads—
one on either end of the socket—to hold it in place for soldering.
Then install the L298 directly to the board. I couldn't find a socket for this chip (see **Figure 4.17**).

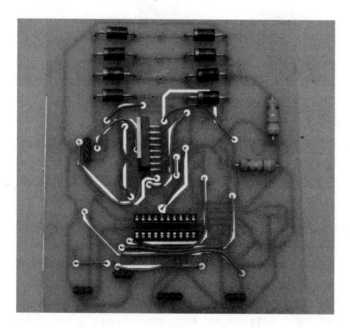

Figure 4.17

Installing IC holder and
L298.

The only capacitor with polarity is the 470UF, so make sure that
this capacitor is installed correctly (see **Figure 4.18**).

Figure 4.18

Correct placement of 470UF capacitor.

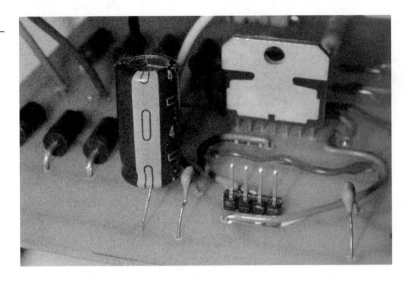

The wire I used for the motor power coming off the board is 20 gauge, cut from the 3-pair shielded cable I am using to connect the motors to the driver boards (see **Figure 4.19**).

Figure 4.19

Spool of cable.

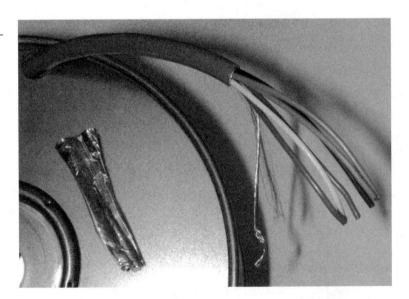

Cut and strip four pieces of wire for each driver and solder them directly to the board (see **Figure 4.20**).

If you want to use a method other than direct soldering, find a suitable connector and install it at the locations of the motor lines.

Figure 4.20

Soldering wire directly to the board.

The L298 needs to have a heat sink installed. I made heat sinks by cutting up a large heat sink I removed from a dead power supply (see **Figure 4.21**). **Figure 4.22** shows the completed driver board.

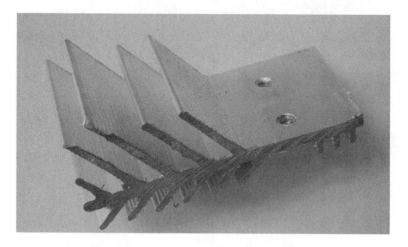

Figure 4.21

Heat sink.

The Interface Board

Refer to the interface figure to determine the component placement. Install the jumper and the resistors first. Cut headers and install them, then install the DB25 connector. This board is used to

Figure 4.22

Finished driver board.

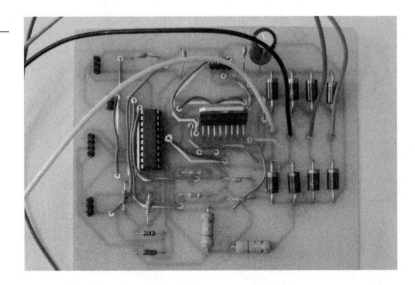

connect the computer to the drivers, the drivers to themselves, and to connect the limit switches to the computer (see **Figure 4.23**).

Figure 4.23

Finished interface board.

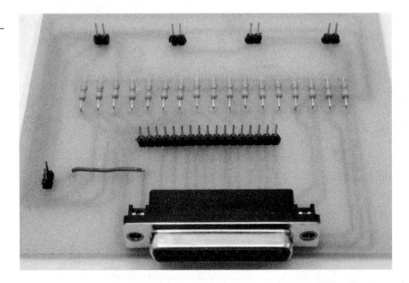

You will now have all four of the printed circuit boards assembled, and if you didn't have much experience soldering before this chapter, you do now. It's nice to have the boards complete and be able to admire your work, but you'll want to find out if they can do their jobs. In the following chapter, you will install KCam 4 and test the boards.

5

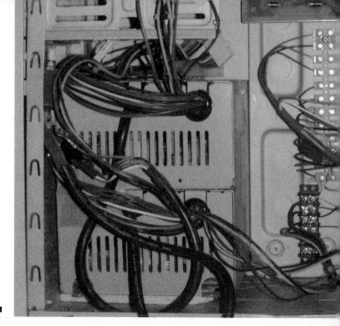

Software Setup and Driver Testing

Material Needed

- Computer

- Windows 95 or 98

- Straight-through parallel cable

- Copy of KCam Version 4

- Finished driver board

- Finished interface board

- Stepper motor

Before you test the driver boards, KCam must be installed on the computer you will use to run the CNC robot. You will want to make sure Windows doesn't need to perform too many tasks while KCam is running, so remove the clock from the status bar at the bottom of the screen or wherever you have it running. Close any programs that run in the background. You can use *control alt delete* to shut them down one at a time. Keep in mind that only two of the background programs need to stay running—*Explorer* and *systray*—all the others can be closed. Go to the KellyWare Web site, www.kellyware.com, to download a demo of the latest version of KCam. At the time I wrote this chapter, the most current was version 4.0.1. Once this is installed, run the program and open the system timing window as seen in **Figure 5.1**.

Figure 5.1

Open Setup then
System Timing.

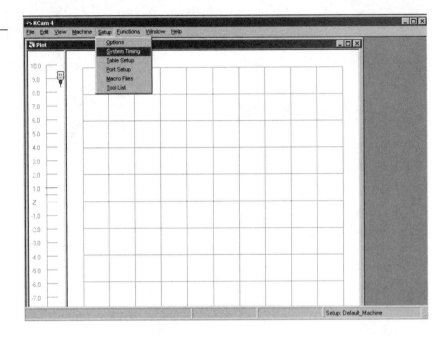

Figure 5.2

Run the system timing
calibrator.

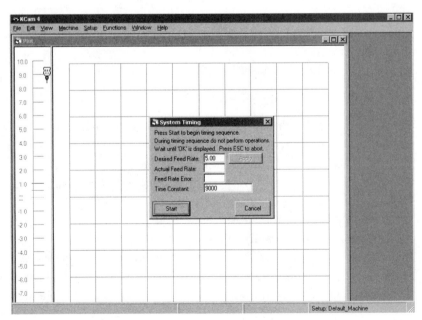

Perform a system timing calibration, use the default feed rate for the calibration, and don't do anything while KCam works (see **Figure** 5.2).

When the timing has been set, the window will look something like **Figure** 5.3.

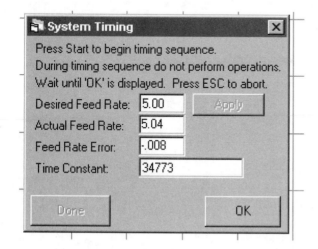

Figure 5.3

Timing finished.

Next, open the *Port Setup* dialog (see **Figure** 5.4).

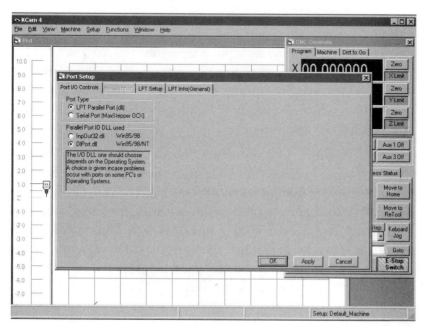

Figure 5.4

Port Setup.

This window allows you to assign parallel port pins to drive the stepper motors. The first tab should be set to LPT parallel port so we can use the printer port. I used the DLL already in use as the default. Next, click on the *LPT setup* tab (see **Figure** 5.5).

Figure 5.5

LPT setup window.

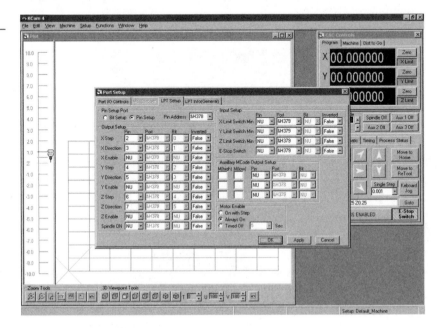

I am using pin 2 for x step and pin 3 for x direction, pin 4 for y step and pin 5 for y direction, pin 6 for z step and pin 7 for z direction. I also set the motor enable setting to "Always On." For testing, you won't need to set any more of the options, so apply the changes and close the window. Next, open the Table Setup window (see **Figure** 5.6).

You will need to tell KCam how many steps the motor needs to receive in order to move your slide or the gantry 1 inch. I am using 2-degree stepper motors, which means it takes 180 steps to turn the shaft 360 degrees. I am using direct drive, so the gear ratio is 1:1 and the lead screw has eight turns per inch. Multiply the number of steps to turn the shaft 360 degrees by the number of turns on an inch of your lead screw. For example, $180 \times 8 = 1440$, so I enter 1440 for each of the axes because all my motors and lead screws are the same. You can set the axis length if you like, but at

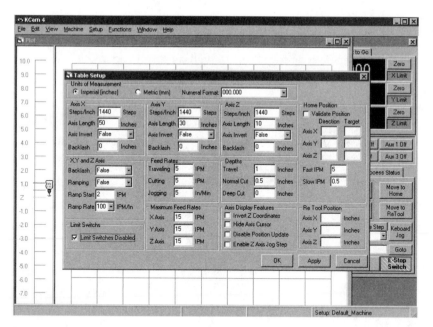

Figure 5.6

Table Setup window.

this point it doesn't matter what you tell KCam. Check the box that indicates that the limit switches are disabled, click apply, and close the window. Next go to the view tab and open the CNC Controls window if it isn't already open (see **Figure** 5.7).

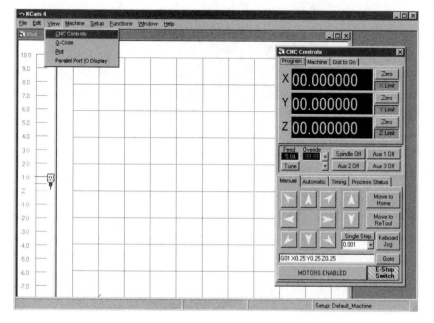

Figure 5.7

CNC Controls window.

You will need to connect the driver boards to the interface board, as indicated in the Port Setup window. Connect all the sync pins on the driver boards together as well as all of the ground wires at header 4 (see **Figure 5.8**).

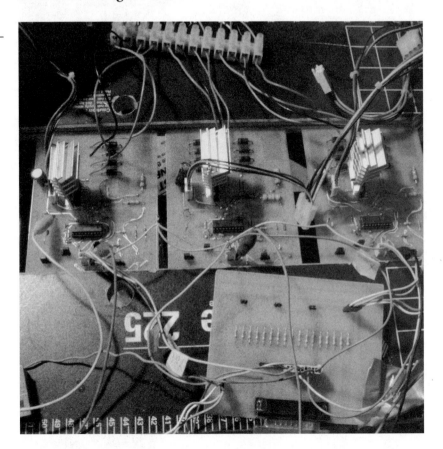

I am using an old PC power supply for each of the driver boards and taking 12 volts for the motors and 5 volts for the circuit. To power the interface board, take 5 volts from one of the boards at the positive side of a capacitor connected to 5 volts and ground (see **Figure 5.9**).

Attach the interface ground to the common ground of the driver boards.

Connect your parallel cable to the printer port of the computer and to the interface board. Now connect a motor to the driver boards.

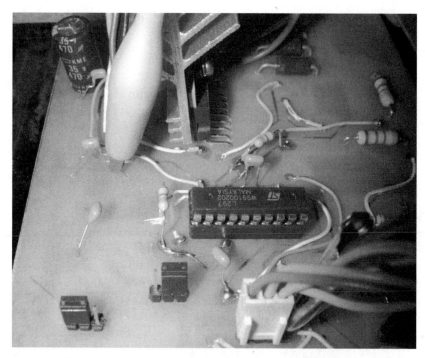

Figure 5.9

Voltage and ground for the interface board.

Carefully verify that you have everything connected properly, especially the motor and circuit voltage wires (see **Figure 5.10**).

If you give the motors 5 volts, nothing bad will happen, but if you accidentally give the circuit 12 volts, the L297 and L298 chips will

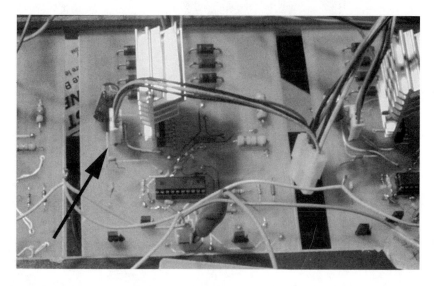

Figure 5.10

Ready for testing.

start to smoke and crackle. I wasn't paying attention late one night while testing some boards and ended up destroying all the chips on three boards. Once you are sure everything is OK, turn on the power supplies. The first indication that the drivers are working and that the motors are good is that the motor shaft will be rigid. It shouldn't be easy to turn the shaft by hand. You will be able to turn the shaft manually a little, but it should take a bit of effort. Next, start moving the motors clockwise and counter-clockwise using the appropriate arrow in the CNC Control window as seen in **Figure** 5.11.

Figure 5.11

Moving motors through CNC Controls window.

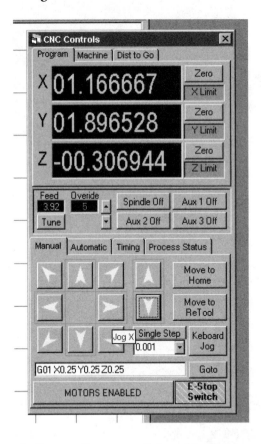

If all is well, you will be able to move all three motors in both directions. If one of the drivers isn't working, check the motor wire connections first to make sure they are in the correct order. Then check the through-holes and traces on the board to verify that all the connections are good. I did have a board that wouldn't work at first. After checking the through-holes from top to bottom of the board, I discovered that three of them weren't actually connected from top to bottom, so I put longer pieces of paper clip through the holes, and the board worked fine.

Creating Test Files

Now that you have all the boards working, open a text editor like Notepad, and type in the following lines of code.

```
N001 %
N002 G90
N003 M03
N004 G00 Z1
N005 G00 X000.000 Y000.000
N006 G00 X002.904 Y002.052
N007 G00 Z0.5
N008 G01 X006.967 Y002.052
N009 G01 X006.967 Y006.115
N010 G01 X002.904 Y006.115
N011 G01 X002.904 Y002.052
N012 G00 Z1
N013 G00 X000.000 Y000.000
N014 M05
N015 M30
```

Save the file as Square-gc.txt. In KCam, go to the file menu and click on "Open G code file". Find Square-gc.txt and open it (see **Figure 5.12**).

The plot view will now be showing a square in red and lines in blue. The red lines indicate where a cut will be made and the blue lines are where the machine will travel without cutting. Click on the automatic tab in the CNC Controls window and run the program you have just written. This square program will move all the motors during its execution. It's a simple test and will indicate that the electronics are finished and can be put in a box. If you want to run a few more tests, the following can be created in Notepad as the first test was and opened in KCam.

Figure 5.12

Opening Square-gc.txt.

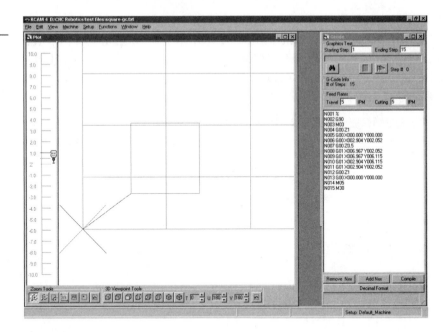

Triangle Test

Use the following code to write a G code file that will plot a triangle.

```
N001 %
N002 G90
N003 M03
N004 G00 Z1
N005 G00 X000.000 Y000.000
N006 G00 X004.112 Y003.411
N007 G00 Z0.5
N008 G01 X007.553 Y003.411
N009 G00 Z1
N010 G01 X007.551 Y003.391
N011 G00 Z0.5
N012 G01 X005.608 Y006.772
N013 G01 X004.113 Y003.411
N014 G00 Z1
N015 G00 X000.000 Y000.000
N016 M05
N017 M30
```

Save this file as Triangle-gc.txt. I include the gc in the name to indicate that it is a G-code file but you could just change the extension of the file to .gc from .txt if you like and you won't have to use the "All Files" filter when looking for the file with KCam (see **Figure 5.13**).

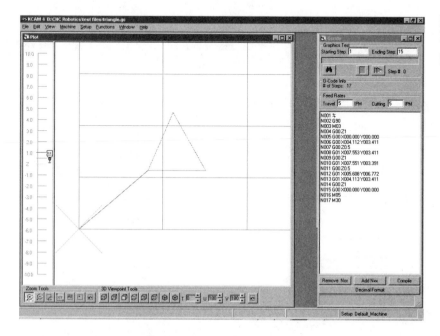

Figure 5.13

Triangle file open in KCam.

Circle Test

Use the following code to write a G-code file that will plot a circle.

```
N001 G90
N002 M03
N003 G00 Z001.000
N004 G00 X1.882 Y2.5
N005 G00 Z000.500
N006 G01 X1.882 Y2.5 Z0.5
N007 G02 X4.118 Y2.5 I003.000 J002.500
N008 G01 X4.118 Y2.5 Z0.5
N009 G02 X1.882 Y2.5 I003.000 J002.500
N010 G00 Z001.000
N011 G00 X000.000 Y000.000
```

N012 M05

N013 M30

Save this text file as Circle-gc.txt and test your motors (see **Figure 5.14**).

After you have successfully tested the drivers and motors using

Figure 5.14

Running the Circle file.

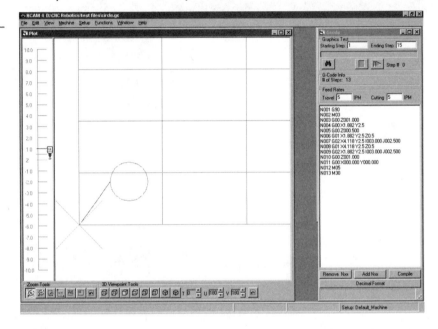

KCam and the files you have written, the electronics can be housed in an enclosure of your choice.

Putting the Electronics in a Case

I am using a box from a computer to house the drivers and the interface board. I bought the case without a power supply for $15 Canadian from a local electronics supply store (**Figure 5.15**).

I removed the power supply holder and the sheet metal used to hold the floppy and hard drives. By removing the unneeded mounting hardware you free up a lot of space. I decided that a fan would be a welcome addition to the box, so I bolted one on the case at the fan opening, as seen in **Figure 5.16**.

Figure 5.15

Computer box.

Figure 5.16

Fan mounted on case.

Next, I drilled three holes in each of the driver boards and corresponding holes on the motherboard mounting surface (see **Figure 5.17**).

Figure 5.17

Holes drilled to mount drivers.

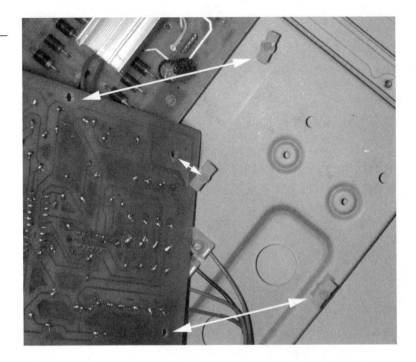

Use standoffs that will snap into place (see **Figure 5.18**).

Figure 5.18

Standoff connector.

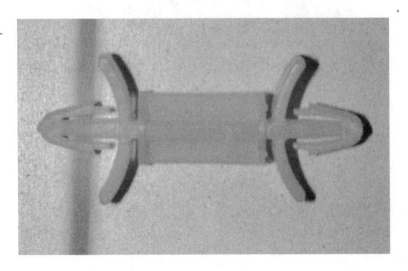

The interface board also has three holes drilled through it and corresponding holes in the floor of the case. The standoffs are high enough to allow the DB25 connector to pass through the bottom expansion slot opening of the computer case (see **Figure 5.19**).

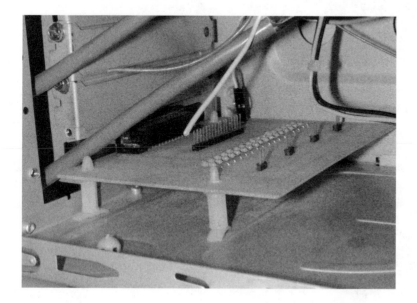

Figure 5.19

Interface board mounted on floor of case.

Next, install a 12-terminal connector strip for motor wire attachment and a 4-terminal connector strip to supply power and ground to the interface board and fan, and to provide a place to connect the ground wires from each of the driver boards (see **Figure 5.20**).

Figure 5.20

12- and 4-terminal connectors.

I used the ground from the 5-volt and 12-volt supply wires just as the power supply connectors are wired (see **Figure 5.21**).

Figure 5.21

4-terminal connector provides 5 and 12 volts.

The next step is to make cables that will be used to connect the driver boards to the interface board. Pinch the required pins onto the wires that will connect the driver's step and direction pins to the interface board at either end. Connect a pin to the driver end of the wire for the sync and ground pins but don't connect a pin to the interface end of these wires. The sync wires can be twisted together and protected with electrical tape. The ground wires can be twisted together and connected to the 5-volt ground terminal on the 4-terminal connector. At the interface end of the wires, place each wire in the connector hole that corresponds to the LPT pin you have assigned to step and direction signals for each axis. Remember that the pins on header 1 of the interface board are numbered 1–18 and are connected to the parallel port pins 1–18 (see **Figure 5.22**). Install your cable when it is finished.

Connect the fan wires to the appropriate voltage on the four-terminal connector. The fan used in this project requires 12 volts.

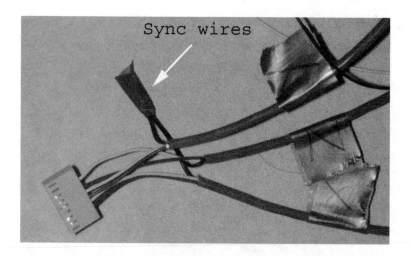

Figure 5.22

Connector at interface end of driver cables.

Use two computer power supplies, one to power the boards and motors on the y- and z-axes, plus another to power the board and motor of the x-axis and the case fan. Because the hard drive and floppy bay were removed, the two power supplies fit nicely in the case, as you can see in **Figure 5.23**.

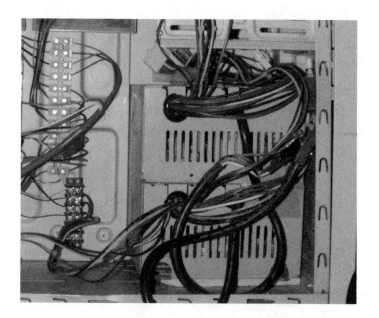

Figure 5.23

Power supplies inside box.

Cut a hole in the panel the power supplies, as in **Figure 5.24**, to allow the power cords to be plugged in. Turn the switches on and tuck the wires you won't use around the power supplies.

129

Figure 5.24

Hole cut in computer
case panel.

Connect the motor wires from the driver boards in the correct
sequence on the 12-terminal connector. Make cables to connect
the motors to the 12-terminal connector with 2- or 3-pair shielded cable. Install the male end of the connector at the motor wires
and the female end to the cable (see **Figure** 5.25).

Figure 5.25

Motor to cable
connection.

Then connect the cables to the corresponding terminal at the 12-terminal connector (see **Figure 5.26**).

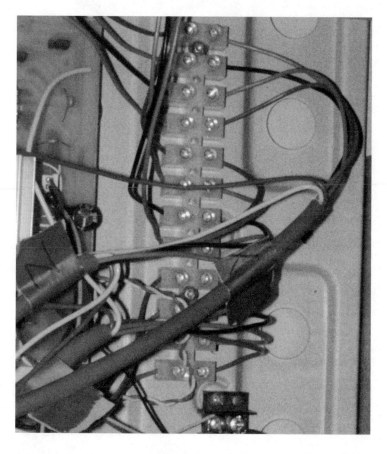

Figure 5.26

Motor cables connected to 12-terminal connector.

Of course, if a motor doesn't turn in the correct direction, just reverse the order of the wires for that motor at the 12-terminal connector for your machine to function properly after the motors are installed (see **Figure 5.27**).

With KCam 4 installed and set up you were able to perform a few successful tests of the drivers and interface boards. After correcting any problems, if there were any, you will have mounted the electronics in some kind of enclosure. The finished product will look tidy and provide some protection for the boards. The electronics are now behind you. In Chapter 6 you will start construction of your CNC machine by building the frame.

Figure 5.27

Completed box.

6

The Frame

Tools and Material

- Guide rail support material

- Cross member material

- Cutoff saw or a hacksaw

- Drill or drill press

- Drill bits

- Nuts and bolts

- Wire brush

- Grinder

- Welder (optional)

- Clamps

- Square

As you recall, I decided on the dimensions of the machine (length and width) largely because of the available space in my work area (see **Figure 6.1**). I really should have built this a little smaller, because as it stands the footprint will dominate my shed and I will

Figure 6.1

Drawing of proposed frame.

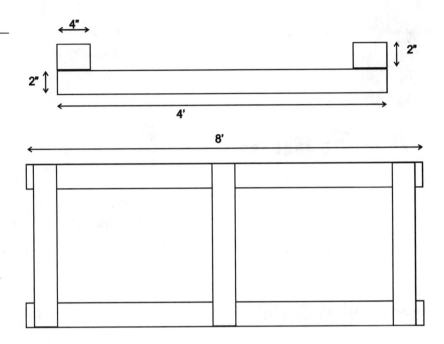

be forced to find alternative storage for a lot of the material and tools for which I don't have an immediate use. But at the current size, I'll still be able to get around it safely to work. Of course, the drawings I made of the machine are really a guide, as opposed to a strict set of instructions. The reason is simple—since I decided on a gantry style of machine, the dimension that I will most closely follow will be the footprint. The material used to construct the machine will determine all the other dimensions. The frame is a good example of this. My drawings use uniform sizes of material in width, height, and thickness, the length being the only variable. I am loath to spend a lot of money on new steel, so the scrap yard closest to my house is my usual first stop. It takes a little more effort to find second-hand material that will work, considering that most of the steel you want is buried under a few tons of steel you don't want.

I was lucky and found the 3.5- × 3.5-inch pieces I am using for the x-axis guide rail supports on top of the pile. The material used as cross members was a little deeper but well worth the effort to dig out. From the 2.5- × 2.5-inch material I used, I cut two pieces 49.5 inches long for either end of the frame and from the 3- × 2-inch material I cut the center cross member. The leftover material from the 3.5- × 3.5-inch material was used to cut the gantry bearing support pieces (see **Figure 6.2**).

Don't forget to bring a tape measure and a straight edge with you to make sure the steel you use isn't warped or bent. The guide rail supports must be straight. If they are not straight, adjusting the rail height along the length will become more difficult. And if the cross members aren't straight, the axis travel will be over an uneven surface, making material positioning very frustrating. Remember that straight is good and rusty doesn't matter.

Figure 6.2

Scrap steel.

A little rust won't pose a problem but does add cleaning work to the project. Take a look at the rusty material I picked up at the scrap yard. The first order of business is to cut the material to length. I have an abrasive chop saw to use for this purpose (see **Figure 6.3**).

Figure 6.3

Steel in saw ready for cut.

The material is heavy enough to tip the saw when clamped in place, so I supported it at the far end. Ideally you want a cut that is square to the sides of the material; as long as you don't use the ends as a mating surface for anything during construction, square cuts don't matter. Having the pieces cut to length, I fit them together upside down on a flat surface, or in my case a really old, wavy, concrete floor (see **Figure 6.4**).

It won't matter if the surface is perfect because when you clamp the pieces together, the frame will straighten out. After the cutting and fitting (to make sure I hadn't made any mistakes), I cleaned the surfaces that would be in contact with one another (see **Figure 6.5**).

Figure 6.4

Fitting steel in place.

Figure 6.5

Contact surface cleaning.

The cleaning was accomplished using a wire brush mounted on my small hand drill, then I went over it again using a small orbital sander with 60-grit paper (see **Figure 6.6**).

The rough ends from cutting were removed with a grinder and a metal file. Keep in mind that you can use whatever tools you have on hand to get the job done. If you only have access to hand tools, then cutting and cleaning will just take more time. When the ends are clean, you have two options: bolt them together or weld them together.

Figure 6.6

Drill and sander with
steel.

Bolting

If you choose to bolt, then start laying out the location of the holes
(see **Figure 6.7**).

Figure 6.7

Lay out the bolt-hole
locations.

The center cross rail holes on the bottom of the support beams will need to be tapped to accept the bolts, or you can drill access holes to enable the installation of the bolts and nuts (see **Figure 6.8**).

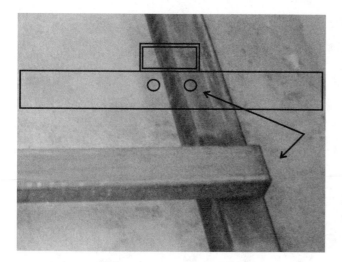

Figure 6.8

Drill access holes. Figure shows the bolt access hole location for center cross member.

Drill all your holes a little oversize for the bolts so that you will have room to adjust the pieces. I like to set up my drill press vise with tape to mark the locations of the holes from the end of the material. In this case, I used 1-1/2 inch centers (see **Figure 6.9**).

Figure 6.9

Using tape to mark hole locations.

I measure the distance from the center of the drill bit to the edge of the vise to get the correct measurement between centers on the width of the piece (see **Figure 6.10**).

Figure 6.10

Setting centers on width of piece.

With a grinder or file, remove the burrs at the bolt holes (see **Figure 6.11**).

Figure 6.11

File off burrs.

Assembly

Put the pieces in place and make sure the two rail supports are parallel and square to each other. Butt the ends against a straight edge to save a little time.

Hold a large square, as in **Figure 6.12**, to check that the ends are square.

Figure 6.12

Large square at work.

Using a good tape measure, check the distance between the pieces at either end, making sure they are parallel (see **Figure 6.13**).

Figure 6.13

Making beams parallel.

Next, place and align the cross members parallel to one another and square to the support beams (see **Figure 6.14**).

Figure 6.14

Cross-member alignment.

Get out your clamps. I have a number of 4-inch C-clamps that will do the job. Make sure you keep checking to see that everything stays in place as you tighten the clamps. If you are bolting, I suggest trying to line up the bolt holes as you align and square the frame, as in **Figures 6.15** and **6.16**.

Figure 6.15

Clamp with bolts.

Figure 6.16

Clamped for welding.

I have an antique arc welder that I picked up at an auction a few years ago. It works just fine, so I chose to weld the frame together (see **Figure 6.17**).

Figure 6.17

Ready for welding.

You aren't going to see a bunch of closeups of my welds because I don't weld very often. They aren't pretty, but they are strong enough to keep everything together. Once the frame is together, clean the rest of the rust off the metal. You can do all this cleaning work prior to assembly; I just like to leave the thorough cleaning until the end (see **Figure 6.18**).

Figure 6.18

Thorough cleaning.

Bearing Rail Support Bolt Holes

I chose to drill the bearing rail support bolt holes before painting so that I wouldn't damage the finished surface. The bearing rails are 6 feet long with holes drilled and tapped on 11-inch centers starting 3 inches from each end (see **Figure 6.19**).

Figure 6.19

Showing the bolt-hole configuration.

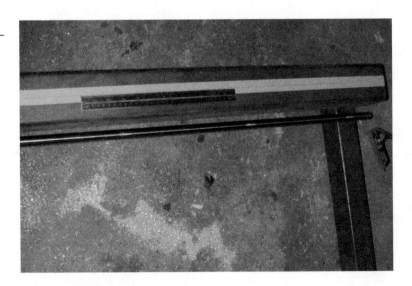

I laid out the holes on a piece of masking tape pulled down the middle of the support rails. I like to use masking tape to lay out drilling locations so the marks are easily visible and quickly changed by reapplying new tape. I found the center of the first support beam and drew a line from end to end using a straight edge (see **Figure 6.20**).

Figure 6.20

Find center.

Starting from a chosen end, I marked the location of each hole at the center line (see **Figure 6.21**).

I then taped the center of the opposing support beam and found the center at the end at which I had started marking the hole locations on the first beam (see **Figure 6.22**).

Figure 6.21

Marking center line.

Figure 6.22

Find center on second beam.

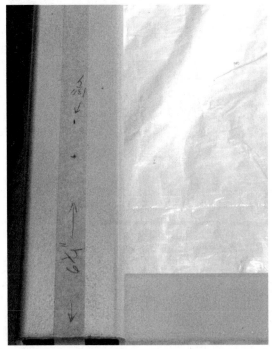

I next measured the distance between the center of the first beam and the center mark I made on the second beam (see **Figure 6.23**).

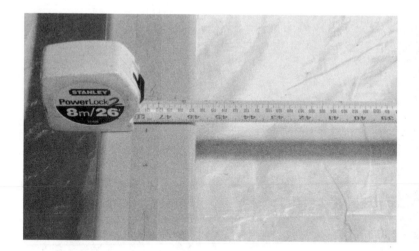

Figure 6.23

Getting measurement.

I used this measurement to mark on the opposite end of the second beam (**Figure 6.24**) and then drew a line with a straight edge, ensuring the two lines were parallel.

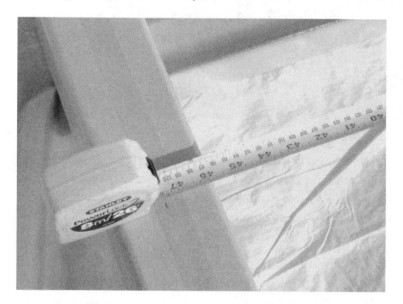

Figure 6.24

Marking second beam for parallel line.

On the second beam, measure the hole locations from the same end of the frame you started from on the first beam. I didn't want to use a handheld drill for the holes, so I rented an electromagnetic drill press, a wonderful tool (see **Figure 6.25**).

Figure 6.25

Electromagnetic drill press.

I set the drill press in place and lined it up, flicked the switch, and it was clamped in place. I wanted the holes to be as square to the top surface as possible because I needed to drill access holes from the bottom of the beam under each rail support bolt hole (see **Figure 6.26**).

After drilling the holes, I put a long bit in the chuck and drilled through the bottom of the beam to create a guide for the large access hole (**Figure 6.27**).

I next flipped the frame over and proceeded to drill the access holes with a 3/4-inch drill bit (see **Figure 6.28**).

Figure 6.26

Drilling holes.

Figure 6.27

Drilling access hole guide.

Figure 6.28

Drilling access hole.

To clean the burrs from the underside of the bolt hole, I put a drill bit in my hand drill and cut the burrs off through the access holes (see **Figure 6.29**).

Figure 6.29

Clean off burrs.

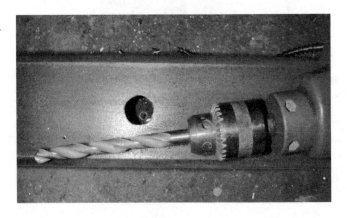

Paint the Frame

Now is a good time to paint. Clean the frame with acetone or something that will remove the rust dust that came from cleaning. If you used new material, then the oil covering the metal will need to be removed. I bought some rust paint at the local hardware store and made the frame pretty, as depicted in **Figure 6.30**.

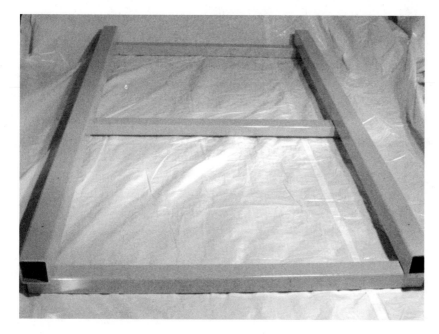

Figure 6.30

Finished frame.

During this chapter you decided how big your CNC machine would be and found the required material to build a frame. After welding or bolting the pieces together, you drilled all the holes needed to install the guide rails for the x-axis and finished the frame with some paint. The final product is sturdy and attractive. During Chapter 7 you will assemble a gantry, install guide rails on the frame, and put the gantry on the guide rails.

7

The Gantry and X-axis

The Gantry

Tools and materials you will require to build the gantry portion of the CNC machine are as follows:

- Two 2 × 4 steel spanning beams

- Two 2 × 6 steel uprights

- Two 3.5 × 3.5 gantry feet

- Drill press

- Abrasive cutoff saw

- Welder

The gantry moves the length of the x-axis and carries the y- and z-axes. It consists of two feet on which the bearing holders are mounted, and two upright posts that have the rail holders for the y-axis spanning between them. The feet are made of 3.5 × 3.5 steel—the same as the frame beams, cut 12 inches long (see **Figure 7.2**).

Figure 7.1

Drawing of proposed gantry.

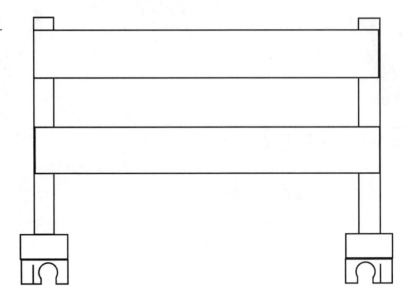

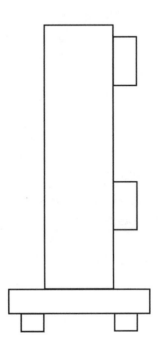

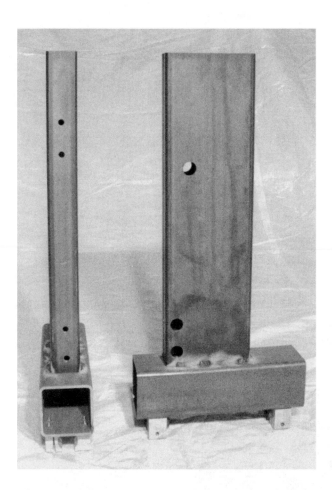

The uprights are 2 × 6 steel cut 20.5 inches long and the spanning beams are 2 × 4 steel, 48 inches long to span the distance
between the two upright posts attached to the gantry feet. These
spanning beams are depicted in **Figure** 7.3.

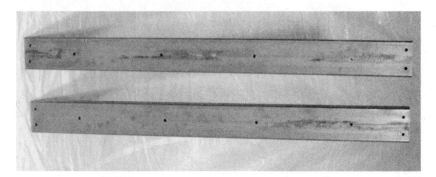

Figure 7.3

Spanning beams.

The spanning 2 × 4s are drilled at either end to accommodate bolts that will be used to attach them to the uprights. The location of the bolt holes is determined by the distance between the centers of the guide rails mounted on the frame beams (see **Figure** 7.4).

Figure 7.4

Finding mounting hole location for spanning beams.

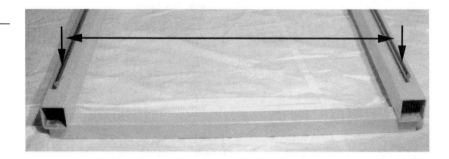

The uprights will need to be drilled to allow the center of each spanning beam to be at the center of the rails that will be installed along their length (see **Figure** 7.5).

Figure 7.5

Location of mounting holes on uprights.

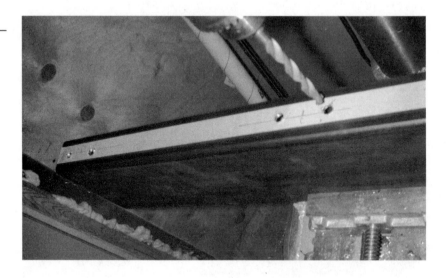

Also drill two access holes on the outside of each of the uprights to allow the installation of the bolts (see **Figure** 7.6).

Figure 7.6

Upright access holes.

Then I drilled holes through the uprights at the location where the lead screw will be inserted. The location of the lead screw holes was determined by the location of the lead screw nut holder on the slide I salvaged from the copy camera (see **Figure** 7.7).

Figure 7.7

Lead screw holes in uprights.

157

I decided to use the existing hole because the nut fit. I need 13 inches from center to center to allow me to use the bearing holders on the platform I removed from the NuArc copy camera (see **Figure 7.8**).

Figure 7.8

Centers of bearing holders determines the bolt hole locations on the uprights.

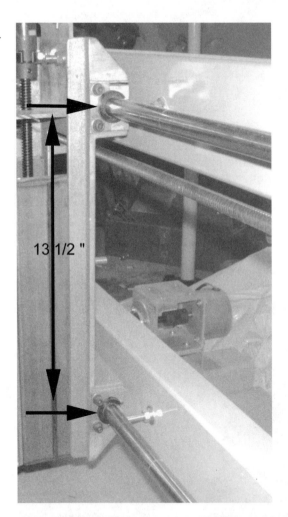

13 1/2 "

I removed the guide rails for the y-axis from the nuArc copy camera for use in my CNC machine. They were longer than I needed, so I cut off the excess with my abrasive cut off saw. The bolt holes along the length of the rails are at 12-inch centers so I drilled four holes in each of the spanning beams directly down the center of their length on 12-inch centers (see **Figure 7.9**).

Figure 7.9

Drilling guide rail
support bolt holes.

Once all the holes were drilled, I clamped the uprights to the feet
with the back of the uprights 2 inches (or the thickness of the
spanning beams) from the end of the feet and as close to center as
possible (see **Figure 7.10**).

Figure 7.10

Clamping uprights to
feet.

I welded the uprights to the feet only on the sides and front where they met the feet. I didn't weld the back, so I could rest the bottom-spanning beam on the foot without any interference (see **Figure 7.11**).

Figure 7.11

Weld locations at gantry feet.

After welding the feet to the uprights, move on to the next section of this chapter and set up the rails on the frame.

The X-axis: Installing the Gantry Bearing Guide Rail

Tools and material required to install the gantry bearing guide rail are as follows:

- Guide rails (**Figure 7.12**)

- Screwdriver

- Wrench

- Nuts washers and bolts

The holes for the guide rail support bolts were drilled during the frame construction process. To be able to install the rail, you must raise the frame to allow an approach to the access holes drilled in the bottom of the rail support beams. I boosted the frame up with a couple of pails I had lying around in my work shed (see **Figure 7.13**).

Figure 7.12

The guide rails.

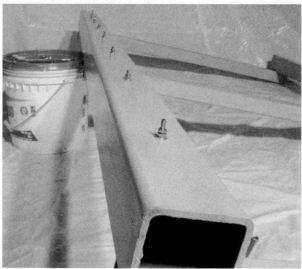

Figure 7.13

The frame supported with pails.

Insert the first and last bolts into the beam through the access holes and turn a couple of nuts onto them (see **Figure** 7.14).

Figure 7.14

Inserting the support bolts.

Screw the first and last bolts into the rail as far as they will go and run one of the nuts to the beam, tightening it to make the bolt rigid. Check the height of the rail at one end and adjust the opposite end to match (see **Figure** 7.15).

Figure 7.15

Finding the working height.

Starting in the middle, install and adjust the remaining bolts alternating from the center. Keep checking with a straight edge to ensure that the rail is straight (see **Figure 7.16**).

Figure 7.16

Checking rail with straight edge.

If the rail isn't straight, you will need to raise or lower a support bolt to straighten it out. Take your time performing this procedure. Follow the same sequence to install the guide rail on the parallel beam, making sure you use the height measurement obtained from the first bolt installed on the previous rail. If this is done correctly, the rails should be parallel and at the same height above the beam, while being straight without any deflection (see **Figure 7.17**).

Figure 7.17

Frame with support rails installed.

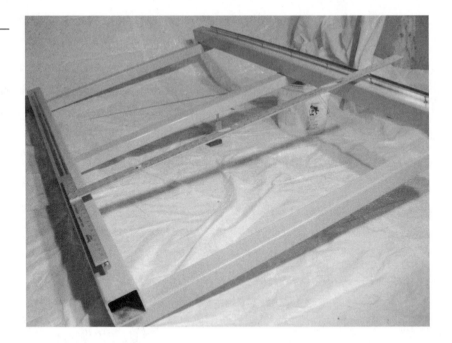

Bearing Holder

Tools and materials required to assemble the bearing holder include:

- Drill press

- Drill bits

- File

- Tap

- 1-1/2 × 3 solid aluminum bar stock

You can buy bearing holders for the linear bearings used in this project from the manufacturers of the bearings, but I decided to make them. They are much less expensive than the manufactured models. I bought a length of 3- × 1.5-inch aluminum bar stock from the metal supermarket. From this stock I cut four pieces 2 inches long (see **Figure 7.19**).

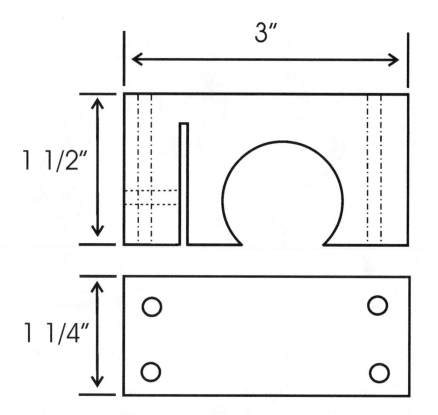

Figure 7.18

Drawing of proposed bearing holder.

3"

1 1/2"

1 1/4"

Figure 7.19

Cutting material.

Next, I drilled a hole in each one, 1-1/4 inches wide to accommodate the outside diameter of the lineal bearings (see **Figure** 7.20).

Figure 7.20

Drilling bearing hole.

I drilled the hole 1/4 inch from the bottom of the stock and then cut out the bottom of the hole for the bolt clearance (see **Figure** 7.21).

Figure 7.21

Cutting out the clearance slot.

At 1-1/4 inch from the side of the hole, I cut a notch in the stock to a 1/2 inch from the top. This notch made the holder adjustable (see **Figure 7.22**).

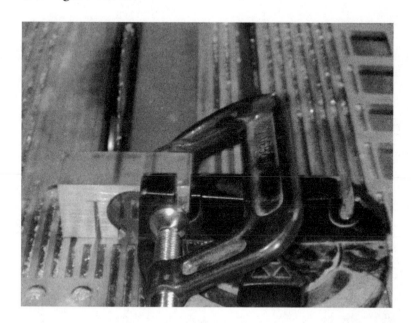

Figure 7.22

Notch cutting.

I drilled four holes from the top down for the bolts that will hold it to the gantry feet (see **Figure 7.23**).

Figure 7.23

Drilling bearing holder installation bolt holes.

The next hole was drilled from the side of the holder until the notch was reached. This hole had to be tapped to accommodate the adjusting screw (see **Figure** 7.24).

Figure 7.24

Tapping hole for adjusting screw.

Because you are working with aluminum, tapping isn't very difficult. When all four holes were finished, I removed any rough edges with a file. Notice that the distance from the top of the bearing hole varies a little from holder to holder (see **Figure** 7.25).

Figure 7.25

Variance between holders.

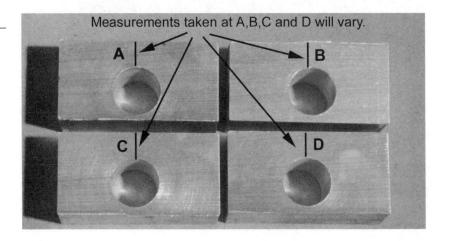

This difference in distance can be removed with sheet metal shims. Because my holders were less than perfect, installing them correctly was critical. I took each of the foot and upright sections of the gantry and turned them upside down, placing the bearing holders on the bottom of the feet with bearings installed (see **Figure 7.26**).

Figure 7.26

Holders in place on bottom of gantry feet.

Ensure that the holders are positioned with the adjusting bolt toward the outer side of the feet and mark each holder with a corresponding mark on the gantry foot. Run one of the rails through the bearings. Tighten the bearing holders enough to remove any play between the rail and bearings (see **Figure 7.27**).

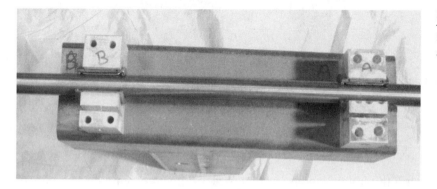

Figure 7.27

Bearing holders adjusted with rail.

Using a measuring device, align the bearing holders with the rail set to the center of the feet (see **Figure** 7.28).

Mark the holes' locations through the bearing holder onto the feet for the mounting bolts (see **Figure** 7.29).

Drill the holes and clean the inside of the feet, removing any burrs with a file.

Place the holders back on the bottom of the feet with loosely attached and determine which of the holders is keeping the rail the farthest from the surface of the foot (see **Figure** 7.30).

Figure 7.30

Finding adjustment height.

This will be the only holder that won't need a shim. Shim the remaining three holders to raise the rail to the height determined from the first holder (see **Figure** 7.31).

Figure 7.31

Using shims to even the bearing holders.

Once the holders are shimmed, tighten the bolts, making sure the rail is directly in the center of the foot. Place each of the feet onto the rail installed on the frame, bolt the bottom-spanning beam onto the upright, and then the top-spanning beam (see **Figure 7.32**).

Figure 7.32

Bolt the spanning beams to the uprights.

Tighten the bolts, ensuring that the beams are square to the uprights and parallel to the bed of the frame. If you can't do both, then it's best that the beam be parallel to the bed of the frame. In any case, the two spanning beams must be parallel to each other, aside from their relationship to the bed or the uprights. Once the spanning beams are installed, the gantry should move easily along the length of the rails. Adjust the bearing holders to remove any excess play but don't tighten them so much that they will bind. If the feet moved freely prior to installing the spanning beams, it is possible that the beams are pushing the uprights too far apart or that once tightened they may cause the feet to toe in or toe out. This can be corrected with shims on one side of a spanning beam (see **Figure 7.33**).

Figure 7.33

Shimming the spanning beams.

With the gantry built and installed on its guide rails, you will want to adjust the gantry's movement using shims and adjusting the bearing holders.

After you have spent a few, patient hours fine-tuning the gantry to enable it to glide along the rails with little effort, you are ready for Chapter 8, the z- and y-axes installation.

8
The Z and Y Axes

The Z-Axis

You'll need the following tools and materials to build the z-axis of the machine.

- 1 × 3 metal stock

- TV sliding tray

- Hacksaw

- Drill

- Drill bits

- Nuts and bolts

The z-axis is constructed from a TV holding glide and two pieces of 1 × 3 foot steel. I bought the TV slide from a big box home improvement store because it looked and felt sturdy enough to act as the z-axis (see **Figure 8.2**).

Figure 8.1

Finished z-axis.

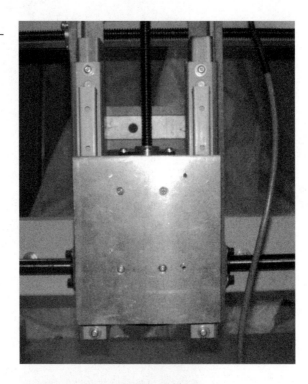

Figure 8.2

TV holder from Home
Depot.

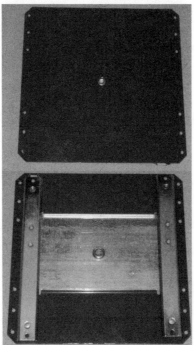

The assembly came with a small swiveling tray attached to the sheet metal spanning the glides. I first drilled the rivet out of the center of the table (see **Figure 8.3**).

Figure 8.3

Drilling out the table's rivet.

Because the y-slide was 5 inches wide inside the outer ridges (see **Figure 8.4**), I needed to cut 4-3/4 inches from the center of the sheet metal holding the slides together (see **Figure 8.5**).

Figure 8.4

Inside dimension of y-slide from NuArc copy camera.

Figure 8.5

Cut 4-3/4 inches from the sheet metal holding the glides together.

4 3/4"

Take off the plugs installed on the slides (see **Figure 8.6**).

Figure 8.6

Remove these plugs.

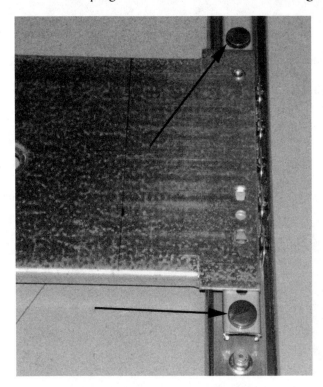

Then cut a piece of aluminum to fit across the width of the slide (see **Figure 8.7**).

Figure 8.7

Aluminum z-slide tool mounting table.

Drill four holes and countersink the top of each hole on the aluminum used for the z-tool mounting surface (see **Figure 8.8**).

Figure 8.8

Drill and countersink table holes.

I drilled two holes on each side of the y-slide with corresponding holes on the 1 × 3 steel used to raise the z-axis assembly from the y-slide surface (see **Figure 8.9**).

Figure 8.9

Mounting holes for the
1 × 3 steel.

I needed room between the bottom of the z-slide and the top of the y-slide as clearance for the acme screw bolt holder to be mounted on the underside of the z-slide. I utilized the holes already present in the y-slide at the bottom for the bearing block needed to hold the acme screw that moves the z-slide up and down (see Figure 8.10).

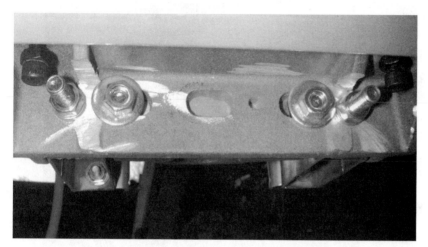

Figure 8.10

Location of z-acme screw support bearing.

The bearing block was too wide to fit between the 1 × 4 z-slide supports, so I trimmed off about 1/2 inch from either side and modified the bolts I used to hold it down (see **Figure 8.11**).

Figure 8.11

Modifications to bearing holder and bolts.

I used the distance from the center of the lower bearing hole to the underside of the z-slide to find the center of the hole I drilled in the angle aluminum that holds the acme screw nut. I then drilled the hole the screw would pass through, and the holes needed to bolt the nut holder to the angle (see **Figure 8.12**).

Figure 8.12

Find the center of acme screw for angle aluminum, A equals B.

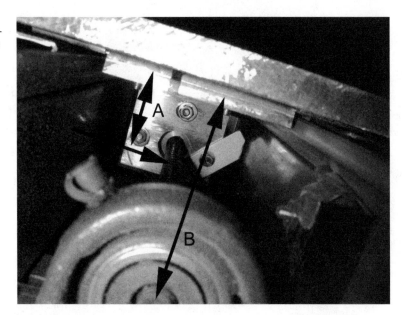

Next I drilled two holes spaced to match the top two holes drilled through the z-faceplate and slide-bearing sheet metal (see **Figure 8.13**).

Figure 8.13

Angle drilled for mounting on underside of z-table.

With the z-slide complete, I placed it on top of the 1 × 3 risers on the y-slide and marked hole locations to match the top and bottom hole locations for the bolts to hold the glides in place (see **Figure 8.14**).

Figure 8.14

Z-axis mounting holes.

I tightened the bottom bolts first, then pushed the faceplate up to align the top of the glides. After aligning the top, I tightened the top bolts. The bolts on the bottom of the z-faceplate could now be tightened, but I left the top ones that also support the acme screw nut holder a little loose, as its position would need to set when the acme screw and stepper motor are installed.

The Y-Axis

In order to assemble the y-axis, you will need the following tools and materials:

- Guide rails

- Slide from NuArc copy camera (or four homemade bearing holders mounted on an aluminum plate)

- Carriage bolts, nuts, and washers

- Wrench

The y-axis essentially consists of the two spanning beams on the gantry, the bearing support rails mounted on them, and a slide from my disassembled copy camera. The bearings in the slide are new, as I was only able to salvage four of the bearings from the camera. Because the camera bearings are the same size as the new ones I bought, they are a perfect fit in the bearing holders on the slide. To install the rails, I first put the carriage bolts in each of the holes for the support rails (see **Figure 8.15**).

I made sure that I didn't run the carriage bolts too far through so they wouldn't interfere when I bolted the rails at either end. I placed the rails through the bearings on the slide (see **Figure 8.16**) and held the top rail in place, so I could screw the bolts at either end of the rail. Next, I screwed in the end bolts on the bottom rail. With the centers of the rails at the required distance from each other, bolting the rail is easy. Remember that the holes you drill need to be a bit bigger than the bolts you're going to use, so if the centers are a little off, it won't matter (see **Figure 8.17**).

Figure 8.15

Z-axis mounting holes.

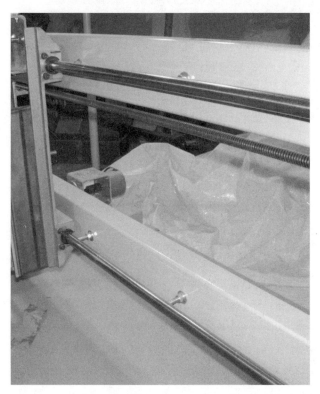

Figure 8.16

Insert the rails through
the bearings.

Figure 8.17

The rail centers should be at the bearing holder centers.

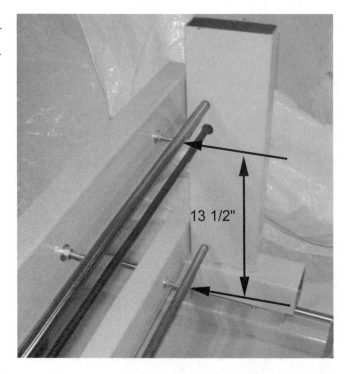

13 1/2"

After the end bolts are screwed in, screw in the rest of the bolts, but don't tighten anything up yet. The rails should be the same distance from the spanning beams and the distance from the spanning beam should allow the lead screw to run through the center of the lead screw nut holder location (see **Figure 8.18**).

Adjust the distance to accommodate the lead screw, then, using a measuring device and a straight edge, bring the rails out to the same distance from the spanning beams and tighten the bolts on either side of the spanning beam to keep them in place (see **Figure 8.19**).

Adjust the bearing holders to remove any slack. Move the y-slide back and forth on the guide rails and make any adjustments necessary to ensure smooth sliding. At this point, you want to be certain that all of your axes move easily and don't bind. You have now completed the installation of the z- and y-axis on the gantry that travels the x-axis. Most of your CNC machine is finished, only lacking stepper motors to give it life. In the next chapter, you will install lead screws and stepper motors to drive each axis.

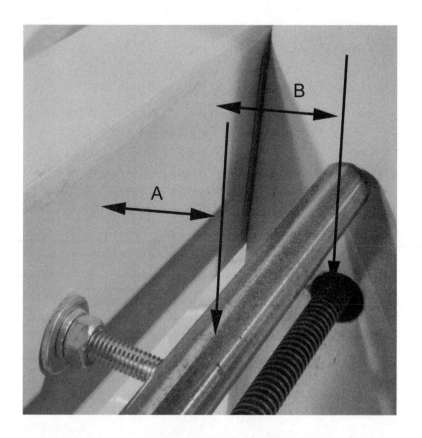

Figure 8.18

Distance from spanning beam allowing lead screw to pass through holes in uprights, A equals B.

Figure 8.19

Use measurement and straight edge to adjust guide rail.

9
Motor and Lead Screw Installation

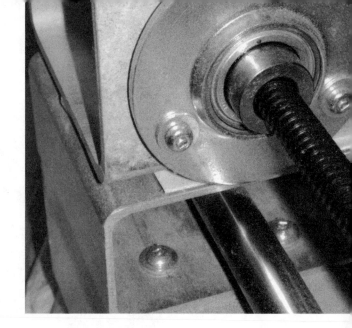

Tools and Material

- Three stepper motors

- Three lengths of lead screw cut to size

- Five pillow blocks with bearings

- Three acme screw nuts

- Three nut holders (only two if you have found a copy camera to cannibalize)

- 3 × 3 aluminum angle iron

- 4 × 4 aluminum square material

- 1.5 × 4 aluminum material

- Drill

- Drill bits

- Nuts, washers, and bolts

You will need to fabricate motor mounts using 4 × 4 aluminum. Cut three pieces 4 inches long (see **Figure 9.1**).

X-axis

The mount used on the x-axis can be drilled in the center for the motor shaft. I drilled a 3/4-inch hole for play and drilled the four holes to bolt the motor to the mount. Drill a 3/4-inch center hole opposite the motor shaft opening and line up a pillow block to mark the holes needed to mount it on the side opposite the motor (see **Figure 9.2**).

I wanted the lead screw to run through the foot of the gantry, so I cut a piece of 1-1/2 × 4 inch aluminum to raise the holder from the beam (see **Figure 9.3**).

Figure 9.2

Drilling holes in motor mount.

Figure 9.3

Spacer for motor mount.

I drilled four holes in the riser from side to side and four holes in the motor mount to correspond to the holes in the riser (see **Figure 9.4**).

Figure 9.4

Holes in riser and motor mount to match.

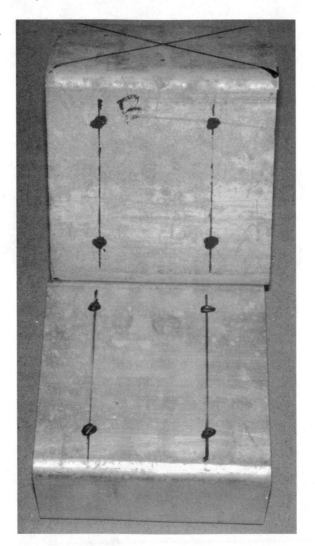

I fastened the riser to the beam using self-tapping screws by inserting a long driver bit through the top holes to engage the self tapper (see **Figure 9.5**).

Figure 9.5

Installing the riser.

After bolting the motor mount to the riser, I measured the distance from the center of the 3/4-inch hole in the mount to the beam (see Figure 9.6).

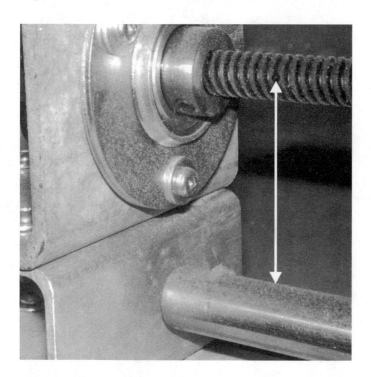

Figure 9.6

Lead screw center measurements.

Cut a piece of angle aluminum on which to mount the acme screw nut. Place the angle aluminum on the foot and mark the center of the lead screw hole using the previous measurement (see **Figure 9.7**).

Figure 9.7

Marking nut holder location on angle mount.

Drill the lead screw hole and holes for mounting the nut holder as well as four holes for bolting the angle aluminum to the gantry foot. Transfer the holes' locations to the gantry foot and drill them out (see **Figure 9.8**).

Figure 9.8

Mounting position of nut holder on gantry foot.

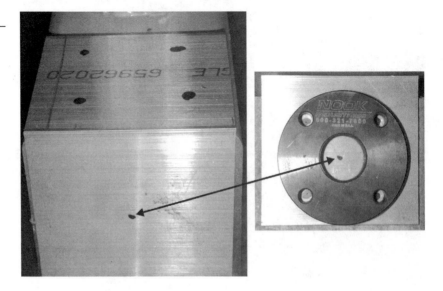

Mount the nut holder on the angle and run a nut onto a 6-foot length of acme screw. Tighten the nut into the holder (see **Figure 9.9**).

Figure 9.9

Acme screw in place.

At the motor end, install a bearing on the outside of the motor mount and push the acme screw into place (see **Figure 9.10**).

Figure 9.10

Installation of bearing at motor mount.

At the non-motor end of the beam, install another riser and the bearing holder with bearing, and insert the lead screw through the bearing (see **Figure 9.11**).

Adjust the length of screw coming through the bearing at the motor mount to allow for the flexible shaft coupling that will connect the motor to the lead screw. Make sure the lead screw is the same height from the beam at either end, and shim it if necessary (see **Figure 9.12**).

Tighten the lead screw to the bearings with the setscrews. Turning the screw by hand should be really easy, without binding. If the screw doesn't turn easily, adjust the position of the lead screw nut at the gantry foot until the screw will turn easily throughout its travel, then tighten the nut holder to the foot (see **Figure 9.13**).

Figure 9.12

Check lead screw
position and shim.

Figure 9.13

Adjust position of nut.

Head back to the motor end and place a flexible coupling on the lead screw (see **Figure 9.14**).

Figure 9.14

Flexible shaft coupling installed.

Insert the motor shaft through the 3/4-inch hole into the flexible coupling and align the motor with the lead screw. Bolt the motor in place and tighten the set screw at the bearing (see **Figure 9.15**).

Figure 9.15

Motor installed.

If you did these steps correctly, the motor should be able to move the gantry from one end of the x-axis to the other without binding. If the lead screw binds, the motor will stall and lose steps so that whatever you are trying to make will not turn out as expected. The step-syn motors are strong enough to run this axis at 10 inches per minute without stalling.

Y-axis

Run a piece of acme screw 55 inches long through one side of the gantry hole and through the nut holder location on the y-slide. Turn a nut onto the acme screw and insert it into the holder location at the y-slide (see **Figure 9.16**).

Figure 9.16

Inserting acme screw.

If your threads match, tighten the nut, or drill a couple of holes beside the nut and insert two screws to keep the nut in place (see **Figure 9.17**).

Figure 9.17

Screws holding nut in place.

Before you send the screw through the next gantry upright, put it through a bearing block (see **Figure 9.18**).

Figure 9.18

Insert lead screw through a bearing block.

Also install a bearing block on the outside of the upright opposite the motor end (see **Figure 9.19**).

Figure 9.19

End of travel bearing block.

I used self-tapping screws to hold the bearing blocks in place. You will need to move the acme screw at either end to find a position that allows the screw to turn easily. I started this quest by checking the distance from the top rail to the top of the screw at the nut (see **Figure 9.20**) and made the ends of the screw at the inside of the uprights the same, but it still took a while to get it properly aligned.

Figure 9.20

Rail-to-screw
measurement.

With the acme screw properly aligned, take another motor mount
and drill out a 3/4-inch hole on two opposing sides. Also drill
mounting holes for the motor that are matched on the opposite
side for self-tappers to fasten the motor mount to the upright (see
Figure 9.21).

Figure 9.21

Holes in mount.

Place the motor mount with the lead screw through the center of the 3/4-inch hole and self tap it to the upright. Place a flexible coupling on the lead screw and install the motor, aligned with the screw (see **Figure 9.22**).

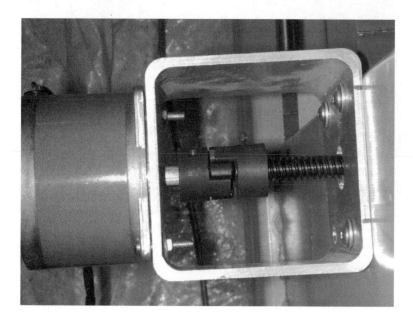

Figure 9.22

Motor installed.

Z-axis

The lead screw bearing is already installed, as are the nut holder and the nut. Cut a piece of acme screw 14 inches long and screw it through the nut and into the bearing. Tighten the bearing set screw to hold onto the lead screw. Drill a 3/4-inch center hole through both sides of a motor mount and drill the motor bolt holes (see **Figure 9.23**).

Figure 9.23

Holes to mount motor.

At the bottom of the motor mount, drill two holes to take advantage of the holes in the y-slide to bolt the mount through (see **Figure 9.24**).

Figure 9.24

Motor mount location on y-slide.

Bolt the mount to the y-slide, making it as square as possible to the lead screw. Insert a flexible coupling on the lead screw and mount the motor aligned with the screw (see **Figure 9.25**).

Figure 9.25

Z motor installed.

Make sure the screw turns easily. When it does, tighten the bolts holding the angle aluminum on which the lead screw nut is mounted (see **Figure 9.26**).

Figure 9.26

Tighten nut holder bolts.

Limit Switch Installation

The tools and materials required to install the limit switches are as follows:

- 6 limit switches

- 1" aluminum angle

- Drill

- Hacksaw

- Nuts and bolts

- One pair shielded cable

- Connectors for limit switches

- Connectors for the interface board

The use of limit switches is optional for this machine, but you can choose to install them as I did. Without limit switches installed, the machine has no way of identifying the boundaries of the usable area of the table. If your machine isn't told to stop, it will continue to travel past the boundaries of the usable area. It won't know that it has stopped moving but the motors will be trying to turn the acme screw and losing steps with every try. This will throw your positioning way off. The machine will not be able to go back to the home position you have established. Of course, leaving the boundaries should only occur if the thing you want to make is wider or longer than your machine can handle. You will be able to do this. See if your project is outside the limits of your table after your file has been imported into KCam. KCam will plot the file over the dimensions you have input pertaining to the size of the x-, y-, and z-axes. As an accuracy precaution, limit switches are beneficial. Also KCam 4.1 will not let the machine manually home the axis if limit switches are not present. I located the limit switches used on this machine at a surplus store. They aren't really microswitches like the kind used in plotters. I think they came out of washing machines or dryers: The first clue was the name Speed Queen on the side of the switch (see **Figures 9.27** and **9.28**).

Figure 9.27

Side one of limit switch.

Figure 9.28

Speed Queen side of
limit switch.

These Speed Queen parts are perfect as limit switches since they
are normally open, and close when the plunger has been pushed
in. Each of the axes requires two switches—one installed at either
end of travel. The switch needs to make contact with part of the
moving axis before the axis gets jammed up.

X-axis Limits

On the x-axis motor end, two holes are needed to bolt the limit
switch to the top of the motor and bearing mount (see **Figure 9.29**).

Figure 9.29

Location of x home limit
switch.

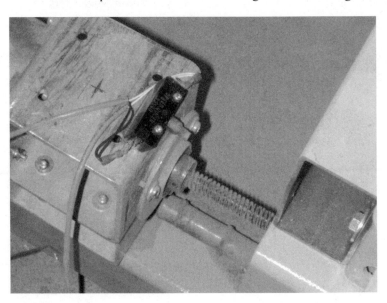

Even though the switch is on a bit of an angle, it still makes contact with the gantry and closes as needed. For the limit at the end of x travel, I made a mount to bolt the switch to; this in turn is screwed to the bearing holder. It is installed on an angle to allow the plunger to make contact with the acme screw nut holder at the front of the gantry foot (see **Figure 9.30**).

Figure 9.30

Limit switch at end of x-axis travel.

Y-axis Limits

The switch at the home position of the y-axis is mounted on a piece of 1-inch angle aluminum that is screwed to the inside of the gantry upright, enabling the plunger to make contact with a bolt protruding from the side of the y-slide (see **Figure 9.31**).

Figure 9.31

Y-axis home position
limit switch location.

The y-axis end-of-travel limit switch is also mounted on an angle bracket, which is screwed to the inside of the upright (see **Figure 9.32**). Self-tapping screws are usually useful here.

Figure 9.32

Switch at end of y-axis
travel.

Z-axis Limits

To install the home position limit switch for the z-axis, a spacer is needed to move the plunger into a position that will allow the acme screw nut holder to make contact. The spacer is 1/2-inch aluminum bar cut the size of the switch body. Drill two holes in the bar that correspond to the holes on the switch body. Next, drill the same hole pattern in the 1 × 3 riser mounted on the y-slide at a position that allows the most z travel (see **Figure 9.33**).

Figure 9.33

Z-axis limit switch at home position.

The end-of-travel limit switch for the z-axis is installed without a spacer because the bearing block mounted on the y-slide won't allow one to be used. Drill two holes in the 1 × 3 to allow the switch to be installed. Additionally, drill a hole above the position the body will be in, so the wire used to attach this limit switch can be fed up through the 1 × 3 to the home limit switch (see **Figure 9.34**).

Figure 9.34

Position of the z-axis
end-of-travel limit
switch.

The problem encountered by mounting the switch directly to the side of the 1 × 3 is that nothing will hit the plunger when the z-axis reaches its end. To resolve this challenge, make an extension from a piece of 1/8-inch thick by 3/4-inch wide aluminum. Drill a hole so that one of the bolts holding the acme nut holder in place can be used to install it. Cut the corner on an angle to give the extension more surface area that can make contact with the plunger of the limit switch (see **Figure 9.35**).

Figure 9.35

Extension to activate
limit switch.

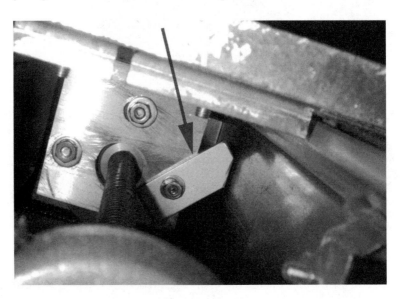

Both of the switches on each of the three axes are wired in parallel so that when either of the plungers is engaged the machine will stop moving. Only one pin of the parallel port is used per axis so the switches need to be wired as in **Figure 9.36**.

Connect to Jp2,3,4 or 5 on interface board

Figure 9.36

Wire the axis limit switches in parallel.

The wire connecting the end-of-travel switch on the x-axis can be fed through the frame beam to the home limit switch and the end-of-travel y-axis switch wire can be fed through the top spanning beam of the gantry to reach the y home switch. At all of the home switches, connect enough wire to the poles to allow the machine to move to any position on the working area of the table without being under stress. The distance of your electronics from the machine will also determine the length of the cable. At the end of each cable install a two-hole connector to fit the header material used on the interface board. Plug each wire into either Jp2, 3, 4, or 5 of the interface board and make sure you open the LPT setup window in KCam so that you can assign the correct pin for each of your limit switch circuits (see **Figure 9.37**).

Figure 9.37

Interface limit switch connection locations.

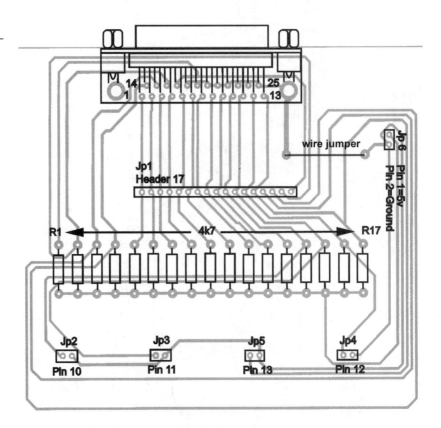

Your CNC machine now has all of the motors, lead screws, and limit switches installed. The Workshop Bot is complete. Chapter 10 will show you what kind of files you'll need to run your machine and how to make them.

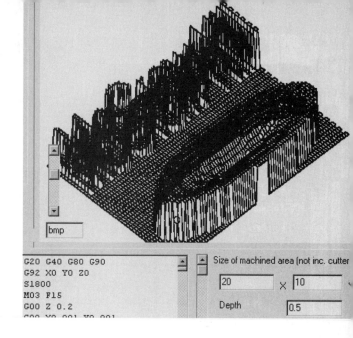

```
G20 G40 G80 G90
G92 X0 Y0 Z0
S1800
M03 F15
G00 Z 0.2
```

Size of machined area (not inc. cutter

20	X	10
Depth		0.5

10

File Creation and KCam

KCam CNC Controller Software

To complete this section of your CNC robotics project, you will need to have KCam installed and configured for your machine. If you haven't already installed KCam, visit the Kellyware Web site at www.kellyware.com and download the demo version of KCam 4. The demo will run fully functional for 60 days, which should be more than enough time to fine-tune your machine setup and start working on all the projects you conceived while building the workshop bot. Chapter 5 has explained setting up KCam in order to test your driver boards. In testing, it really didn't matter if all the parameters of your machine had been established, but it now does. Open KCam, click on *Setup* and open the Table Setup window. Ensure that the number of steps per inch and the physical dimension for each axis on your machine are correct. Next, uncheck the *Limit Switches Disabled* button, click *Apply* and close the window (see **Figure 10.1**).

Figure 10.1

Table Setup window.

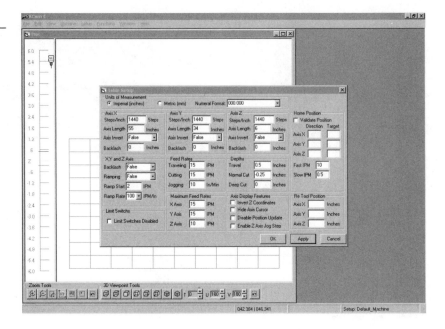

If you are using surplus stepper motors like the step-syn motors on this machine, the maximum rate of travel for each axis will probably be around 15 inches per minute. You will need to experiment to discover the fastest rate of travel for your machine. When you are running the machine too fast, the stepper motors will start to skip turns. With steppers, the faster you run them, the lower their torque becomes and the load they can move decreases dramatically. Keep in mind that every time you change the rate of travel you need to run the system timing utility. Next, click on setup and open the *Port Setup* window. Choose which pins you are assigning to each of the drivers for their step and direction signals and which pins you are using for the limit switch circuits (see **Figure 10.2**).

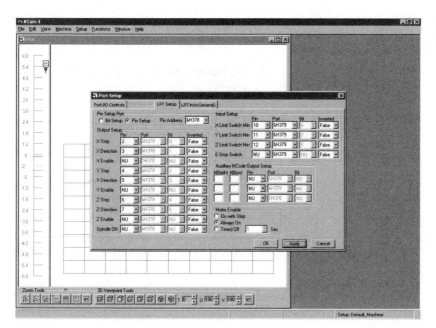

Figure 10.2

Port Setup window.

Apply the settings and close the window. Run the system timing utility again by clicking on *Setup* and opening the *System Timing* window. Click on *Start* and wait for it to finish before doing anything else (see **Figure 10.3**).

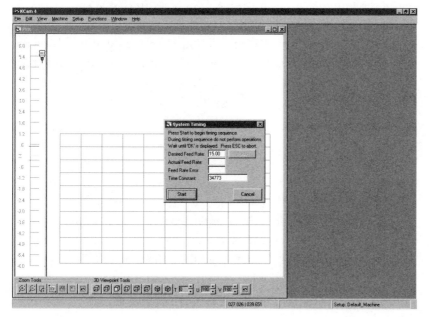

Figure 10.3

System Timing window.

Click on the *View* button and open the *CNC Controls* window (see
Figure 10.4).

Figure 10.4

CNC Controls window.

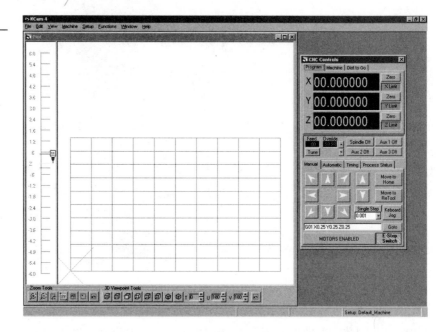

Begin moving the x- and y-axes of your machine and note the x in
the *Plot* window. The x indicates where on the table the center of
the cutting tool or pen is located. The z position is indicated by the
image of a cutting tool that raises and lowers as you jog the axis.
You want your machine's x- and y-axes to move in the same direction as the x in the plot window and your z-axis to move up and
down as the tool image (see **Figure 10.5**).

If you find that the machine is moving in the opposite direction,
reverse the order of the motor wires connected to your driver
board. To determine the exact size of the machine's working area,
move each axis to its home position and zero out that position in
the *CNC Controls* window. Now move each axis in turn to the end-
of-travel switch. The length of each axis will be displayed in the
corresponding distance-traveled box. You may not want to hit
your limit switches every time you home the machine; in this case,
set the zero position 1/4-inch before the home position limit
switch. You can set home or the zero position anywhere you pre-

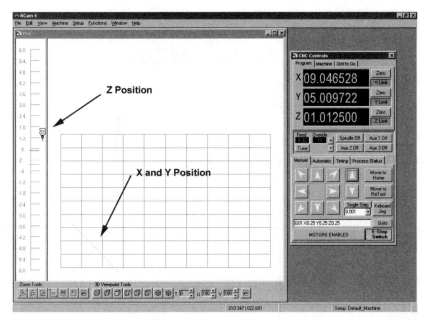

Figure 10.5

Plot window indicates position on the table.

fer on the table, but KCam will assume that the table remains defined by the dimensions you entered in the Table Setup window and that the tool is at the lower left corner. This is handy if you want to make a couple of something from one piece of material without rewriting the G-code or generating another file that has multiple parts to import. Just jog above the first series of cuts, zero the machine, and run the program again.

KCam File Requirements

Now that your machine is working, you will need to generate files that will be opened or imported by KCam. KCam will open G-code files directly and import DXF, Exellon, HPGL, or Gerber file types (see **Figure 10.6**).

Let's start with the G-code. When KCam imports a file, the program translates and generates the G-code that represents the image of the imported file. It all boils down to G-code. So what is G-code? Simply put, G-code is a set of commands that tell the control software where to send the tool and what to do when it gets there. After interpreting the G-code, the software will send

Figure 10.6

Files KCam can work with.

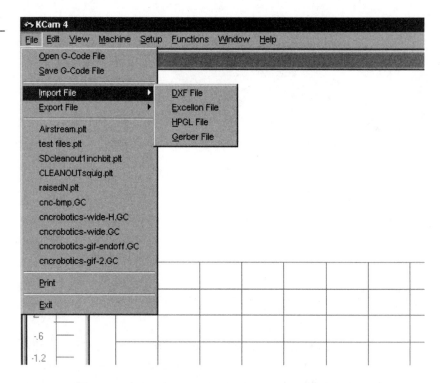

signals to each axis to move in the required direction as many steps as are required to travel the distance. A G-code file can be written in a text editor or in the G-code window in KCam. Aside from G-code, there are also M-code functions. M-code is used to control the behavior of the program and machine rather than controlling the axis movement. **Tables 10.1, 10.2,** and **10.3** list G-code and M-code commands that KCam can understand.

Table 10.1

G-Code Commands

G-code	Description
G00	Rapid Traverse
G01	Normal Traverse
G02	CW Arc
G03	CCW Arc
G04	Execute Dwell Time
G17	XY Plane Selection

(continued on next page)

G-code	Description	Table 10.1
G18	XZ Plane Selection	G-Code Commands (continued)
G19	YZ Plane Selection	
G40	Cancel Cutter Diameter Compensation	
G41	Start Cutter Diameter Compensation Left	
G42	Start Cutter Diameter Compensation Right	
G45	Normal Traverse	
G73	Drill Cycle	
G80	End Drill Cycle	
G81	Drill Cycle	
G82	Drill Cycle with Dwell	
G83	Drill Cycle	
G90	Sets Absolute Mode	
G91	Sets Incremental Mode	
Pxx	Sets Dwell Time to xx	
Fxx	Sets Feed Rate to xx	

M-code	Description	Table 10.2
M00	Program Stop	M-Code Commands
M01	Optional Program Stop	
M02	Program End	
M03	Engage Spindle Relay	
M04	Engage Spindle Relay	
M05	Disengage Spindle Relay	
M06	Tool Change	
M07	Mist Coolant On	
M08	Flood Coolant On	
M09	Mist and Flood Coolant On	
M13	Engage Spindle and Coolant	

(continued on next page)

Table 10.2	M-code	Description
M-Code Commands (continued)	M30	Program End and Reset
	M60	Program Stop
	M98	Call Macro subroutine (not available in KCam 4.0)

Table 10.3	M-code	Description
User-defined M-codes	Mxx	Engage user-defined output
	Mxx	Disengage user-defined output
	Mxx	Engage user-defined output
	Mxx	Disengage user-defined output
	Mxx	Engage user-defined output
	Mxx	Disengage user-defined output

Note: "xx" represents the number specified by the user.

If you want to know more about the process of writing G-code, I suggest a trip to the local library or bookstore; pick up a volume dedicated to CNC programming. Most of these G-code and M-code commands are self-explanatory and you will learn more about how they are useful to you as you experiment. The best way to understand how the G-code works is to create a file in a drawing program and import it into KCam, then go through the code and see how it is put together. That brings us to the import function of KCam.

How to Create a File to Import

The two file formats I have been generating are HPGL and DXF; the former is more frequently used than the latter. HPGL is an abbreviation of *Hewlett-Packard Graphics Language*, a command language to control plotters and printers. An HPGL file is only two-dimensional, and while KCam imports this file type it uses the information in the Table Setup window to generate the cutting depth and travel position for the z-axis. KCam will use this information every time a file is imported, so if you find that a z position is either too deep or too shallow, the cutting or traveling depth in the Table Setup window will need to be altered and the file reimported. The changes made to the table setup will not be represented in a G-code file displayed in the editor window—the file *must* be reimported for the G-code to change. A DXF file is a *Data Exchange File* created by the software publisher AutoDesk for AutoCAD software. DXF files are also two-dimensional graphics files supported by virtually all CAD (computer aided design) programs. I have been using CorelDraw to make my files. CorelDraw is easy to use and will export your drawings as either HPGL or DXF.

CorelDraw

You need to gain access to a version of CorelDraw by purchasing the software or by using a friend's computer on which it is installed. Open CorelDraw and create a new graphic (see **Figure 10.7**).

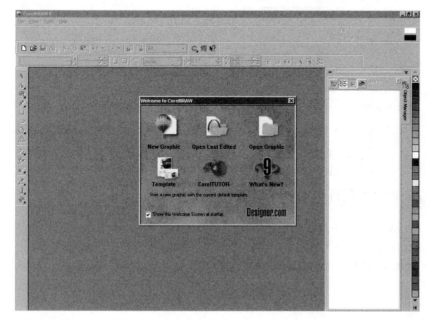

Figure 10.7.

Create a new graphic in CorelDraw.

The default page dimensions are letter size (8.5 × 11-inches) in portrait layout. Let's say you want to make a sign with letters and numbers and the sign should be 10 × 20-inches. Go to the *Layout* button in CorelDraw and from the dropdown list click *Page Setup* (see **Figure 10.8**).

Figure 10.8

Layout options.

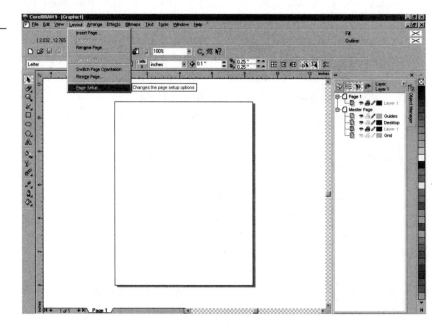

When the *Options* window opens, change the layout to *landscape* and the width and height dimensions to 20 and 10 inches (see **Figure 10.9**).

Now the page reflects the size of the sign that will be made. You will find that when you import into KCam an HPGL file created with CorelDraw, KCam plots the objects you place on the page *from last created to first*. This will affect how long it will take your machine to run through the program. For a first test, place parallel lines across the page, all starting from the left horizontally to the right of the page, stacked from the bottom to the top of the page (see **Figure 10.10**).

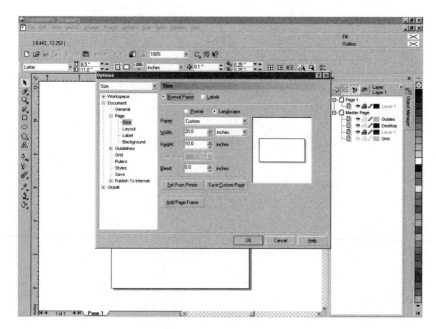

Figure 10.9

Page options window.

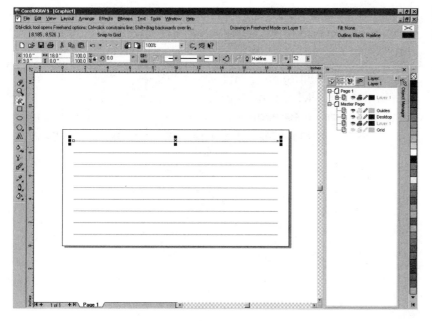

Figure 10.10

Drawing parallel lines.

Save this graphic as *linetest.plt* by exporting it as an HPLT file. To export from CorelDraw open the *File* dropdown menu and click *Export* (see **Figure 10.11**).

Figure 10.11

Click *Export*.

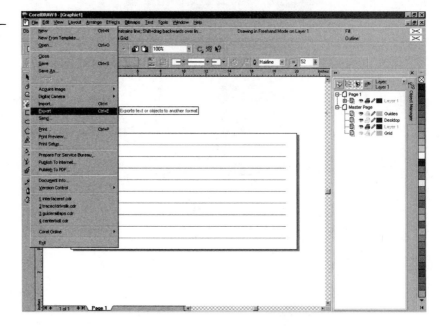

When the *Export* window opens, name the file and choose *PLT-HPGL Plotter File* as file format, and click the *Export* button (see **Figure 10.12**).

A new window will open, containing HPGL export options. Ignore the *Pen* tab and click on the *Page* tab. Make sure this window is set up as in **Figure 10.13** with *Scale* at 100%, *Plotter Origin* as bottom left, *Orientation* as landscape, and verify that the page size is correct.

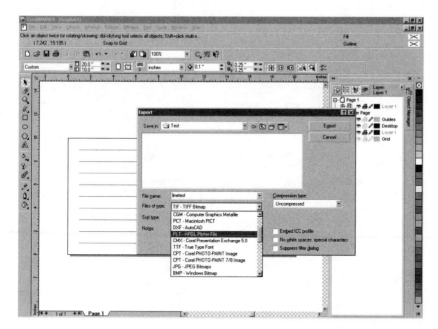

Figure 10.12

Export window.

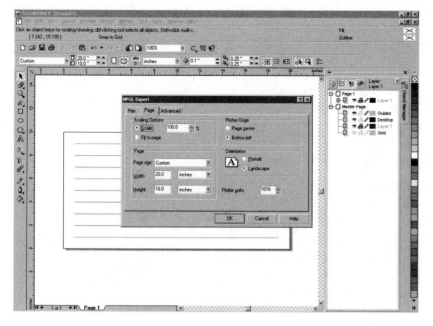

Figure 10.13

HPGL Export, page setup.

Next, click the *Advanced* tab, confirm that *Simulated Fill* is set to *None* (see **Figure 10.14**), and click on the OK button.

Figure 10.14

HPGL export, *Advanced* window.

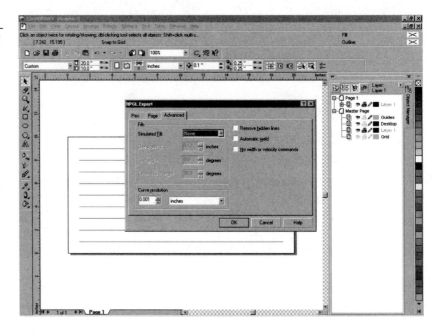

You have now exported the *linetest.plt* graphic as a plotter file. Open KCam and import *linetest.plt* using the *Import HPGL* button. After KCam has plotted the program the G-code will look something like this.

```
N000 [KCam Conversion]
N001 [Original File: linetest.plt]
N002 %
N003 G90
N004 M03
N005 G00 Z0.5
N006 G00 X000.000 Y000.000
N007 G01 X001.016 Y009.144
N008 G00 Z-0.25
N009 G01 X019.304 Y009.144
N010 G00 Z0.5
N011 G01 X001.016 Y008.128
```

N012 G00 Z-0.25
N013 G01 X019.304 Y008.128
N014 G00 Z0.5
N015 G01 X001.016 Y007.112
N016 G00 Z-0.25
N017 G01 X019.304 Y007.112
N018 G00 Z0.5
N019 G01 X001.016 Y006.096
N020 G00 Z-0.25
N021 G01 X019.304 Y006.096
N022 G00 Z0.5
N023 G01 X001.016 Y005.080
N024 G00 Z-0.25
N025 G01 X019.304 Y005.080
N026 G00 Z0.5
N027 G01 X001.016 Y004.064
N028 G00 Z-0.25
N029 G01 X019.304 Y004.064
N030 G00 Z0.5
N031 G01 X001.016 Y003.048
N032 G00 Z-0.25
N033 G01 X019.304 Y003.048
N034 G00 Z0.5
N035 G01 X001.016 Y002.032
N036 G00 Z-0.25
N037 G01 X019.304 Y002.032
N038 G00 Z0.5
N039 G01 X001.016 Y001.016
N040 G00 Z-0.25
N041 G01 X019.304 Y001.016
N042 G00 Z0.5
N043 G00 X000.000 Y000.000
N044 M05
N045 M30

Every x- and y-axis coordinate after a z-axis move to 0.5 inch is a travel move without cutting because the tool is 1/2-inch above the surface of the material, assuming the surface is at 0.0 and every move following a z-axis move to –0.25 is a cutting or plotting move with the tool at 1/4-inch below the surface of the material. With this in mind, notice how KCam goes about plotting these lines. KCam starts at the beginning of the last line drawn on the page and proceeds to travel back to the beginning of every previous line before it drops the z-axis down to cut. This doubles the amount of time it would take your machine to complete the program. Considering all the wasted travel time, see the plot of *linetest.plt* in **Figure 10.15**.

Figure 10.15

How KCam plots *linetest.plt*.

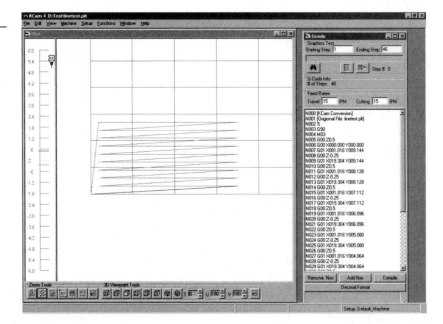

When KCam imports a file, it plots cutting lines as red and travel lines as blue in the plot window. But when an HPGL file is imported, most of the travel lines are also red. To change the travel lines from red to blue, use the G-code window and edit G01 to G00 at every x and y coordinate command following a z move to travel depth. After you are finished editing, click the *Compile* button at the bottom of the *Editor* window and the plot will be regenerated with travel lines blue. Of course, this isn't a problem if the G-code file is small, but it becomes very impractical if the file consists of thousands of lines. The color of the lines plotted in the plot window won't affect the actual movements of the machine needed to create what the file represents, it's just nice to have a well-defined graphical representation of the movements the machine will make so that you can better generate files. Next, open CorelDraw again and create another graphic called *linetest2*; but this time, start drawing the lines from the top right corner across the page to the left and start the next line from the side of the page at which you ended the previous line. Save this file as *linetest2.plt* with the same HPGL options as the last file. Import *linetest2.plt* into KCam and the G-code will look like this:

```
N000 [KCam Conversion]
N001 [Original File: linetest2.plt]
N002 %
N003 G90
N004 M03
N005 G00 Z0.5
N006 G00 X000.000 Y000.000
N007 G00 X001.016 Y000.000
N008 G00 Z-0.25
N009 G01 X019.304 Y000.000
N010 G00 Z0.5
N011 G00 X019.304 Y001.016
N012 G00 Z-0.25
N013 G01 X001.016 Y001.016
N014 G00 Z0.5
N015 G00 X001.016 Y002.032
N016 G00 Z-0.25
N017 G01 X019.304 Y002.032
N018 G00 Z0.5
N019 G00 X019.304 Y003.048
N020 G00 Z-0.25
N021 G01 X001.016 Y003.048
N022 G00 Z0.5
N023 G00 X001.016 Y004.064
N024 G00 Z-0.25
N025 G01 X019.304 Y004.064
N026 G00 Z0.5
N027 G00 X019.304 Y005.080
```

N028 G00 Z-0.25
N029 G01 X001.016 Y005.080
N030 G00 Z0.5
N031 G00 X001.016 Y006.096
N032 G00 Z-0.25
N033 G01 X019.304 Y006.096
N034 G00 Z0.5
N035 G00 X019.304 Y007.112
N036 G00 Z-0.25
N037 G01 X001.016 Y007.112
N038 G00 Z0.5
N039 G00 X001.016 Y008.128
N040 G00 Z-0.25
N041 G01 X019.304 Y008.128
N042 G00 Z0.5
N043 G00 X019.304 Y009.144
N044 G00 Z-0.25
N045 G01 X001.016 Y009.144
N046 G00 Z0.5
N047 G00 X000.000 Y000.000
N048 M05
N049 M30

The lines plotted in the plot window will look like **Figure 10.16**.

Notice the difference in travel. How you draw in CorelDraw should be considered if you want to maximize the efficiency of your machine's movements. Let's go back into CorelDraw, and using the same size page, type some numbers and letters. Type "McGraw-Hill 1 2 3" or whatever else you choose (see **Figure 10.17**).

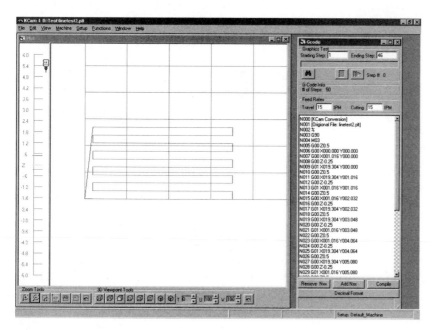

Figure 10.16

KCam plot of
linetest2.plt.

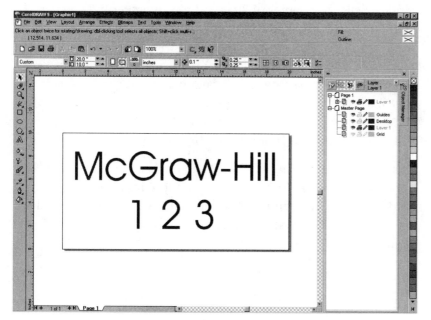

Figure 10.17

Text and numbers in
CorelDraw.

Export this file as *texttest.plt* and import it into KCam. This file will plot as seen in **Figure 10.18**.

Figure 10.18

"McGraw-Hill 1 2 3" in plot window.

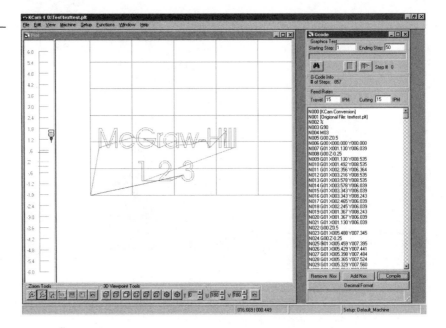

This experiment gives you an idea of how the file would work with various tools. The center of all your tools, whether they be pens or router bits, will follow directly down the middle of the cutting path displayed in the plot window. Here's how to see what results when you use a pen or router bit with a 1/4-inch diameter. In KCam *Setup*, select *Tool list* from the dropdown menu. Double click *Tool #001* and set the diameter to .25. Click OK and set the tool length at .125. This number shows up as the *offset*, which is normally the radius of the tool (see **Figure 10.19**).

Click OK and close the window. When asked if you wish to *Save Default.tol*, click *Yes*. To be able to call a tool from the list, you will need to go to *Setup* and click *Options*. In the first window, titled *General*, make sure that the *Force Plot Bit Radius* is unchecked (see **Figure 10.20**).

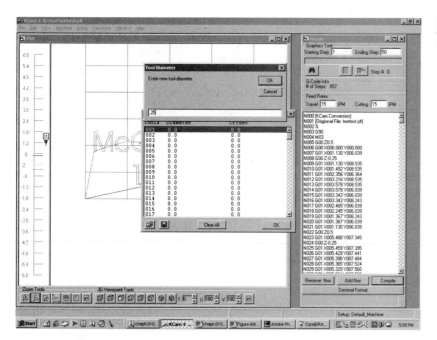

Figure 10.19

Setting tool diameter.

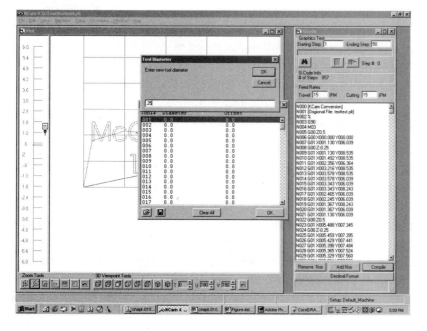

Figure 10.20

Uncheck the *Force Plot Bit Radius*.

Next, click *Compile* at the bottom of the G-code editor window, and your plot should look like **Figure 10.21**, which reflects how a 1/4-inch bit would make the letters appear.

Figure 10.21

Using a 1/4-inch tool.

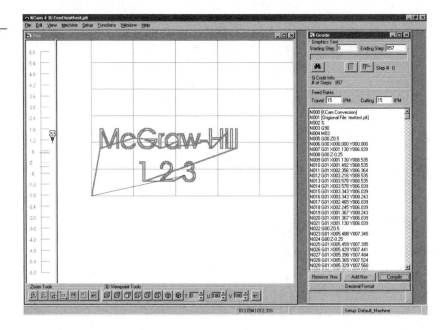

If you leave the first tool set to 0.0 and edit tools from 2 up, you will see thin lines plotted when you first import your graphic, but you can preview a new tool diameter by adding T002 or T003 etc. as a new line in the G-code editor. If you use this command at the beginning of the program (after the G90 command) the entire plot will reflect the tool. If you add it halfway through, then only the lines after the command will reflect the change (see **Figure 10.22**).

Changing bit size in the plot window is a great way to see what a bit or pen will make if it follows the tool path. It is essential to consider the type and size of tool when generating the artwork for a project. You probably have plenty of fonts on your computer with which to experiment and if you don't like the way a font works as a plotter file, then import the plotter file back into CorelDraw and proceed to edit it. Any graphic you create with CorelDraw can also be exported as a DXF file, but before you can export as a DXF you need to convert all the text to curves. Select all the text and open

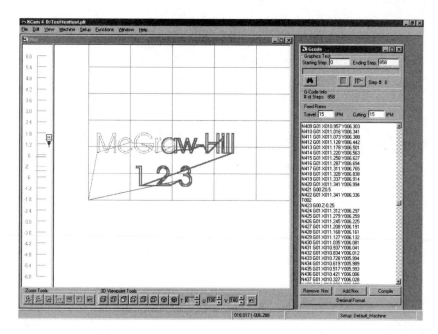

Figure 10.22

Changing bit size for part of graphic.

the *Arrange* dropdown menu to choose *Convert to Curves* (see **Figure 10.23**).

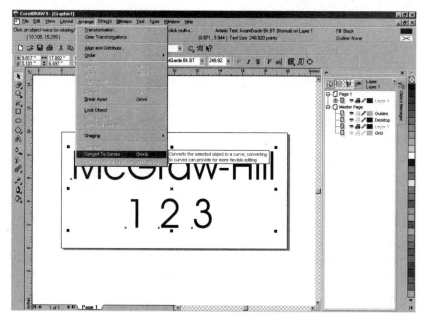

Figure 10.23

Converting text to curves.

Once converted, the file can be exported as a DXF file. The only problem with the DXF file format that CorelDraw creates is that the 0 point is in the middle of the page. When KCam opens it, the G-code takes three quarters of the drawing outside the machine's working area (see **Figure** 10.24).

Figure 10.24

CorelDraw generated DXF file opened in KCam.

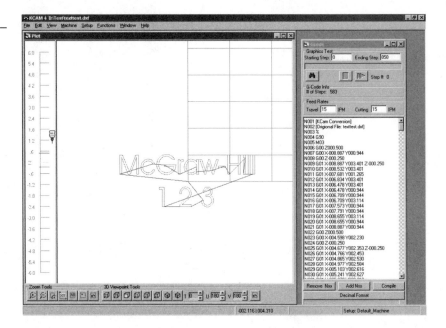

Fortunately, if this happens you can offset all the G-code from within KCam to bring the plot within the table. To do this, you will need to move the y-axis position half of the width if your original graphic and move the x-axis over half of the length so that y=5 and x=10. Open the *Function* dropdown menu and click the *Offset G-code* button. Enter 10 for x-axis, 5 for the y-axis and leave the z-axis at zero. This will bring the plot back to where you want it (see **Figure** 10.25).

Another point to consider is that CorelDraw produces vector artwork. This format can be imported into your graphic and scaled without any loss of detail. Raster images, on the other hand, are bitmaps composed of pixels, which means that as you increase the size of a bitmap you begin to lose detail. If you want to use bitmaps, then import them into a graphic, create a new layer and

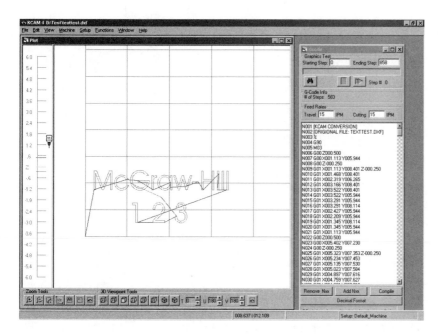

Figure 10.25

DXF after G-code offset in KCam.

trace over it. Once you are finished tracing, delete the imported image and save the vector trace.

ACME Profiler

ACME Profiler–Coyote Edition, Version 6.0.0.0. is software that is able to generate a G-code program that will carve material by raising and lowering the z-axis as the x- and y-axes are moving (see **Figure 10.26**).

KCam will support a more three-dimensional cutting path than our previous examples, but only if G-code is generated with a different program like the ACME Profiler6. You can find it on the Internet at the Simtel software repository. The page where you can download the program is www.simtel.net/pub/pd/60491.html. It comes in a registered version for $20 U.S., or the shareware version (with some limitations) is free. This software is published by Science Specialists, Inc. and their Web site address is www2.fwi.com/∼ kimble/ scispec/scispec.htm. Profiler is a straightforward program to learn and use. It looks at the gray scale of a bitmap image and generates the cutting path by assigning the maximum depth you input to the

Figure 10.26

ACME Profiler6.

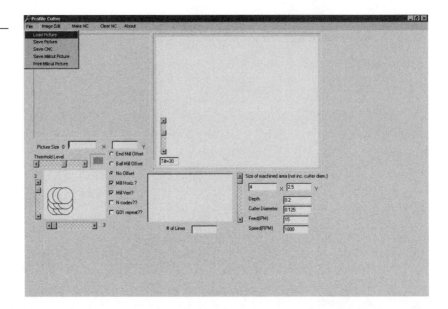

darkest areas, with the cut depth diminishing as it moves to white. Any jpeg or bmp image file will work, but gray-scale images will take less time to load and process. When creating a graphic to open in Profiler, think about depth of cut as shades of grey, then export it as a bmp (Windows bitmap) (see **Figure** 10.27).

Figure 10.27

Creating a graphic for ACME Profiler6.

Load the bmp file into Profiler by clicking *Load Picture* from the *File* dropdown menu. Set the number of overlapping passes you want using the two sliding buttons at the lower left of the screen. Indicate whether you want the tool paths to cut on the x-axis or y-axis, or both. You can have Profiler offset your G-code to compensate for the tool you are using. Tell Profiler the size of the area that you wish to mill. Set the depth of cut you want to make and the cutter diameter. The *Feed* and *Speed* options can be disregarded. When finished, click the *Make NC* button. I used a 1/2-inch cutter diameter for this test to keep the lines of G-code to a minimum. See the result in **Figure 10.28**.

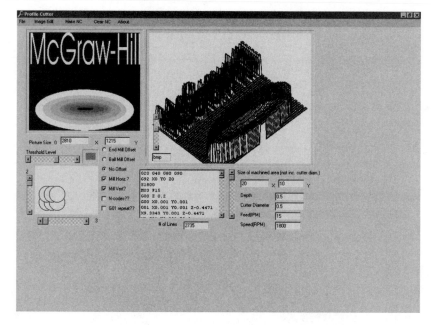

Figure 10.28

ACME Profiler6 generates G-code.

Open the *File* dropdown menu and chose the *Save CNC* button, name your file with a .GC extension so that KCam has no trouble recognizing it. To use this file in KCam, open the *File* dropdown menu and click the *Open G-Code File* button, locate your file and select it, then click *Open*. **Figure 10.29** depicts how the file *profiler6.gc* looks from the top. An isometric view of the plot is shown in **Figure 10.30**.

Figure 10.29

Top view of *profiler6.gc* in KCam.

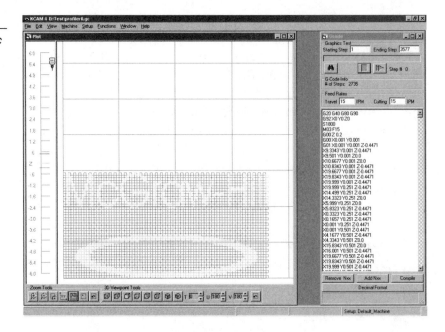

Figure 10.30

Isometric view of *profiler6.gc* plot.

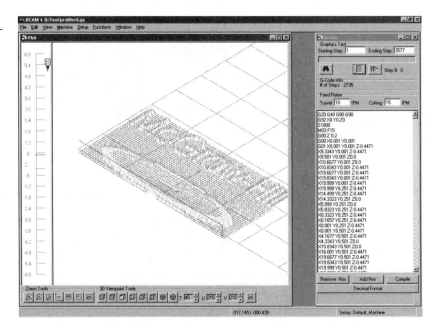

You may not want to use a large bit to cut this kind of profile—a smaller bit and more passes would result in much finer work. You should be able to keep busy generating all kinds of projects for your CNC machine with these techniques. As far as file creation is concerned, however, I have only scratched the surface, so further research on your own will produce even more interesting results. I show you the tool holders I made and tests I performed in the following chapter.

11

Tool Holders and Testing

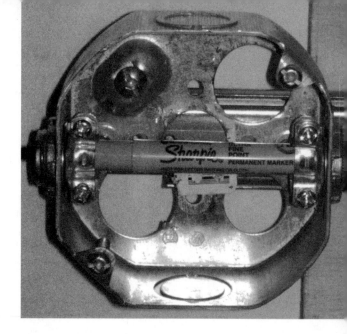

Tool Holders

To build your CNC machine tool holders, you will require the following tools and materials:

- Plywood

- Self-tapping screws

- Large pipe clamps

- Springs

- Drawer slide

- Electrical junction box

- Two electric wire straight-relief clamps

For testing purposes, I built a couple of makeshift tool holders from material I already had in my collection of "might be useful one day" stuff. Neither of these holders positions the tool with great accuracy but for what they are you will be impressed with the results. After you have your machine up and running, you might consider designing better holders. We will build a penholder and a router/dremel holder—we'll make the penholder first. The best and safest way to test the machine is to use it as a plotter. Files that were created during the file creation chapter can be used to test the machine. Create as many test files as you like.

Penholder Tool

The penholder is made with a piece of plywood cut to fit the size of the z-axis table. Drill some holes in the plywood and corresponding holes in the z-table. I drilled the holes in the z-table a little smaller than the self-tapping screws I used. The self-tapping screws work well and eliminate hand-tapping the holes on the z-table. The drawer slide is mounted on the plywood with two small screws, as it would be if used in a drawer (see **Figure 11.1**).

Figure 11.1

Slide mounted on plywood.

Next, take a junction box like the one in **Figure 11.2** and knock out two of the openings that are across from each other (see **Figure 11.3**).

Figure 11.2

Junction box.

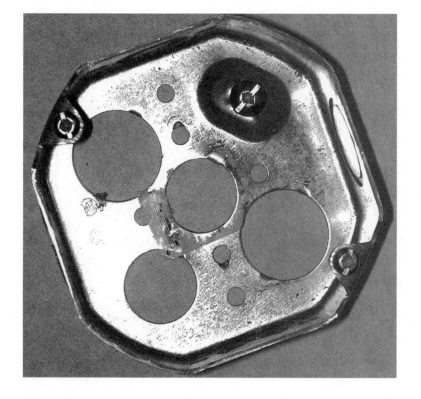

Figure 11.3

Knock out two
openings.

You may need either to cut some of the slide off or drill new holes to mount the junction box as seen in **Figure 11.4**.

Figure 11.4

Cut off some of the slide.

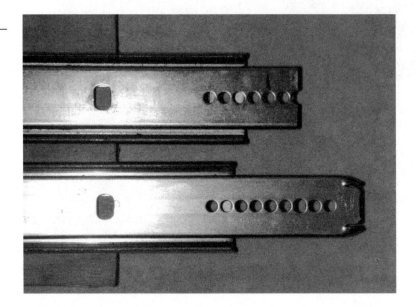

I cut off a little of the end so that I wouldn't have to drill any new holes. Now mount the junction box at the end of the drawer slide (see **Figure 11.5**).

Figure 11.5

Box installed on slide.

Place one of the strain-relief clamps through the top opening of the junction box and one in the bottom opening. Insert them from the inside of the box with the adjusting screws facing out (see Figure 11.6).

Figure 11.6

Installing the strain-relief clamps.

You now need a couple of weak springs to provide tension for the slide. The tension will allow the slide to come back to a home position as well as provide a little pressure for the pen as it travels over the work area (see **Figure 11.7**).

The main objective of the springs is to maintain consistent positioning for travel. If the springs aren't used, the slide may not come back to its original position or it will not drop down to make contact with the paper on the work area. **WARNING: Don't use strong springs—the pen will not last if too much pressure is applied while it is plotting.** See the weak springs installed in Figure 11.8.

FIGURE 11.7

Weak springs.

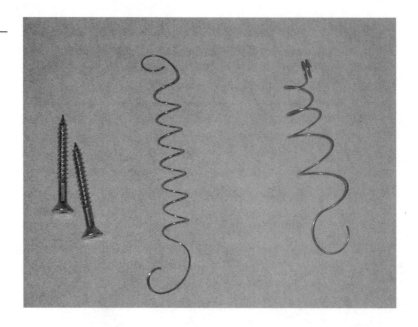

Figure 11.8

Springs installed with screws.

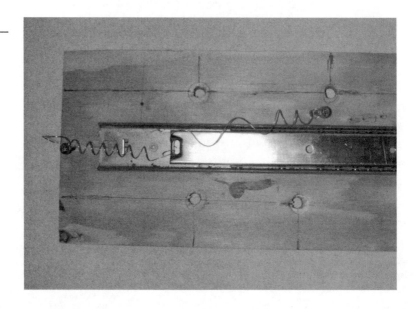

To use this penholder, slide the pen of choice through the strain-relief clamps and tighten their screws, keeping the pen tip below the holder (see **Figure 11.9**).

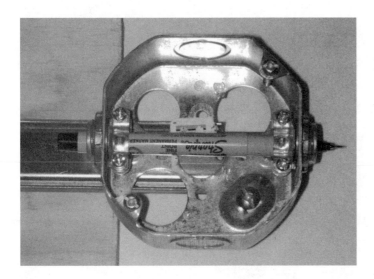

Figure 11.9

Putting pen in holder.

Screw the holder to the z-axis table and use a square to align the slide with the tabletop (see **Figure 11.10**).

Figure 11.10

Use a square to align the holder.

The pen will be almost perpendicular—if the pen is on a severe angle because its body is very tapered, add tape to the narrow area to act as a shim.

Router/Dremel Holder

The router holder consists of two pieces of plywood cut the width of the z-axis table. The back piece will require holes drilled through it for mounting on the z-axis (see **Figure 11.11**).

Figure 11.11

Back of router holder with mounting holes.

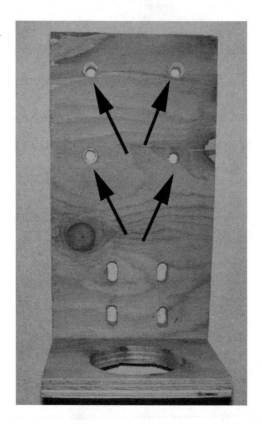

The other piece is cut a little longer than the body of a router and a hole a little smaller than the diameter of the router body is cut 1 inch from the side (see **Figure 11.12**).

Figure 11.12

Base of holder.

On the back plywood, drill or cut two sets of openings with the openings of each set 1-1/2 inches apart and centered on the board. Cut the first set 1 inch from the bottom of the backboard and the next set 2 inches above the first set (see **Figure 11.13**).

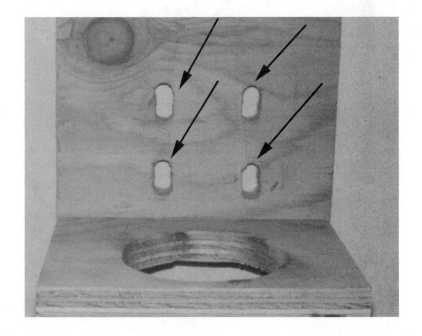

Figure 11.13

Clamp openings cut through back.

On the back of the back plate, remove enough material from between the sets of openings to allow the pipe clamps to recess, keeping the holder flat to the z-table (see **Figure 11.14**).

Figure 11.14

Remove wood between openings.

Unscrew the clamps so that they can be fed from the front through an opening and up through the next opening (see **Figure 11.15**).

Screw the base piece of plywood to the bottom of the back ply. To mount a router, place a piece of wood at the openings between the clamps to move the router out closer to the center of the hole in the base (see **Figure 11.16**).

Figure 11.15

Feeding the clamps.

Figure 11.16

Wood spacer.

Use a piece of wood at each clamp so that the router body doesn't sustain any damage when the clamps are tightened (see **Figure 11.17**).

Figure 11.17

Wood to protect the router.

Tighten the clamps and you're done! To get the router almost perpendicular, place a long bit in the collet and use a square to find an accurate position. The hole in the base of the holder is just large enough to allow the bit to be changed without removing the router (see **Figure 11.18**).

Figure 11.18

Big enough for tool change.

Testing the CNC Machine

Your machine will need a flat smooth surface between the frame beams. If you haven't yet placed anything between the beams, now is the time. I used two sheets of 3/4-inch MDF as the table surface (see **Figure 11.19**).

The sheets of MDF were cut to length and width on the panel saw at the lumberyard where I purchased them. MDF is heavy and dense, but one sheet will probably bend a little between the supports, so I used two. I haven't noticed any deflection yet, although my method of checking is with a straight edge—not the most accurate method. With the working surface in place, install the penholder (with pen) on the z-axis table. Then place a piece of paper over the working area of the table (see **Figure 11.20**).

Figure 11.19

Two sheets of MDF as a tabletop.

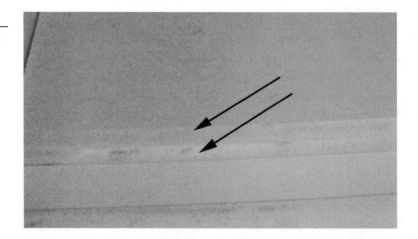

Lower the pen by jogging the z-axis from the *CNC Controls* window in KCam until it reaches the surface of the paper, then allow the z-axis to lower a little bit more to make sure the pen will make contact with the paper even if the surface of the work area isn't perfectly flat. Use this distance as the *default cutting depth* in the *Table Setup* window and have it travel 1/4 inch above the surface of the paper. If you have installed limit switches and have decided where the home position is for all three axes, then just save the setup for the plotter tool as a *KCam machine setup file* so that you won't have to do this again. You can use any kind of paper you want. I have

Figure 11.20

Paper tablecloth taped to work surface.

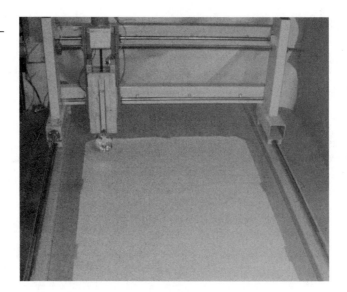

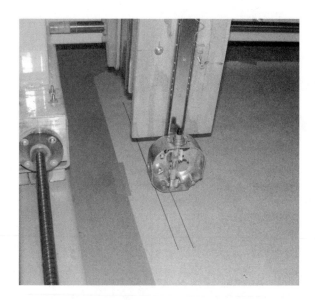

Figure 11.21

Plotting the second line.

a roll of kraft paper on which I initially experimented, but I also purchased a roll of white tablecloth paper from a party supply store. Tablecloth paper is 3 feet wide and 50 feet long, which makes it ideal as plotter paper. Whatever you use will need to be taped at the corners to stop it from sliding around the table. Start with the parallel line plotter file named *linetest2.plt*, created in the last chapter. Import it into KCam and open the CNC Controls window, clicking the *Automatic* tab. Press the *Start* button and watch the machine draw lines. **Figures II.2I** and **II.22** show the machine plotting *linetest2.plt*.

Figure 11.22

Plotting the last line.

The next test will use the file *testtext.plt*, created in the last chapter. I wanted to keep using the paper already taped to the table, so I used the *Offset G-Code* function to move the plot 11 inches up the y-axis (see **Figure 11.23**).

Figure 11.23

Text test plot moved 11 inches along y-axis.

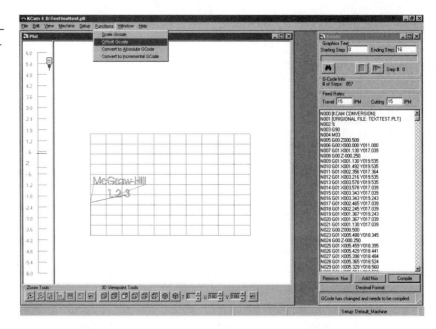

Figure 11.24

Progress shot of texttest.plt plot.

The finished plot is shown in **Figure 11.25**.

Figure 11.25

Text test plotted.

I created one more test file with CorelDraw and exported it as *shapes.plt* (see **Figure 11.26**).

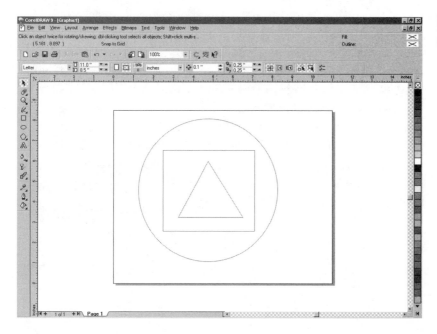

Figure 11.26

Shapes graphic.

Once imported into KCam, this file will need to have the G-code offset to use the same paper. I offset the x-axis 21 inches to put it in front of the line test. The finished plot of *shapes.plt* is shown in **Figure 11.27**.

Figure 11.27

Shapes plot finished.

WARNING: It is wise to test your graphics files with a pen before you start cutting material that costs more than paper. Even if you have no interest in plotting all your files with a pen, it would be wise to run the programs through without a tool on the z-axis to ensure the machine will perform as expected. With your Workshop Bot running smoothly and doing whatever you ask of it, your only concern will be finding as many uses for it as you can. The next chapter deals with examples of what can be done with this machine. There are undoubtedly a variety of uses for this machine other than what I have used as examples in the next chapter—the kind of materials and tools that can be used are only limited by your willingness to learn and your imagination.

12

Examples

This chapter on examples was the most fun to write! Nothing compares to having a new tool to get your imagination working. Every new tool inspires the act of creation, and a home-built CNC machine is no exception! In my opinion, a tool you built from scratch has an appeal unlike any store-bought tool. It's just a whole lot of fun to make things. So that's what I've been doing since completing this project—thinking of numerous uses for my new tool. The uses I offer as examples are certainly not the only functions this machine can perform, though. I'm sure you'll have many more uses and projects for your machine. The rest of this chapter is divided into sections based on the tool I have mounted on the z-axis of my CNC machine. The sections will detail how the machine can be used as a plotter, with a router or Dremel tool, as well as with an engraving tool.

Plotter

As a plotter, this machine can write on almost any surface and because of the large working area, it can perform printing jobs that a standard sized printer cannot. The first plotting job for which I used the machine was a stained glass pattern I created some time ago but hadn't been able to reproduce on a single, large piece of paper. The pattern, inspired by a Frank Lloyd Wright design,

Figure 12.1

A stained glass pattern plotted.

measures 28 × 38 inches to fit in my front door, as shown in **Figure 12.1**.

I used CorelDraw to create the pattern but had only been able to print it on a series of letter-size pieces of paper that I taped together. Take it from me—taping small pieces of paper to form a larger drawing doesn't work very well. I exported the CorelDraw pattern as an HPGL file for KCam to import.

The next two examples are of a WMF (Windows Meta File) format vector drawing that I downloaded from a Russian Web site that has free vector clip-art available. The address is www.clip-art.ru/indexx.html. The first image is of a bulldog, plotted very small (see **Figure 12.2**) and the next image is the bulldog plotted at a large size (see **Figure 12.3**).

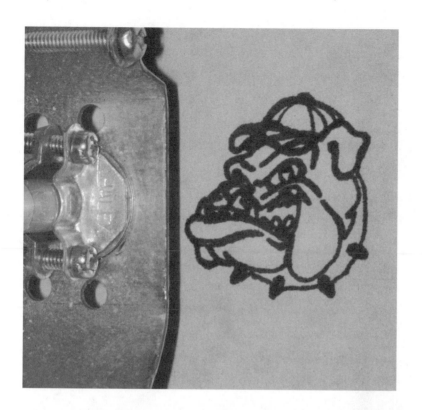

Figure 12.2

Small bulldog plot.

Figure 12.3

Large bulldog plot.

As you can see from the small and large plots of the bulldog, vector images can be enlarged or reduced without loss of detail. I used the same pen for both plots, which resulted in the smaller bulldog plot's having pen lines that bleed into each other. This gives the final plot a little less resolution. Had I used a pen with a smaller tip, gauging the tip to the size of the plot, the two images would look exactly alike.

My next plotting task was to find a pen that could be used to plot on glass surfaces, with ink that would act as a resist to glass etching paste. It took a while to find such a pen. I bought every permanent marker available and tested them by drawing a line on a piece of glass with each and applying etching paste to each line, letting the paste sit for about 15 minutes and washing it off in soapy water. The only pen that worked well as an etching resist was the Pilot Extra Fine Metallic Ink Marker. I plotted a dolphin on a piece of clear glass using the Pilot Metallic Marker. Then I added ink to the perimeter of the drawing with a wider version of the same pen so that the etching paste wouldn't stray from the area I wanted to etch, as you can see in **Figure 12.4**.

Figure 12.4

Dolphin plotted with metallic ink.

Make sure that the ink has been able to dry completely before using the etching paste, or the paste and ink will just mix and your glass etching will not be successful. Be careful to read the safety warnings on the etching paste, as it is a caustic substance. I applied the paste with a small piece of scrap wood, but you could use a brush, as long as metal doesn't make contact with the paste (see **Figure** 12.5).

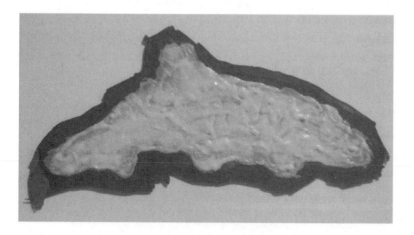

Figure 12.5

Etching paste applied to glass.

I let the paste sit on the glass for about 15 minutes, but the time required to etch sufficiently could differ, depending on the brand of etching paste and the temperature of your work area. **Figure** 12.6 shows the final result after washing the paste off and removing the metallic ink with lacquer thinner.

Figure 12.6

Dolphin etched in glass.

Using the CNC machine as a plotter allows you to generate full-size drawings of patterns for woodworking projects, as it allowed me to plot a stained glass design on a single large piece of paper. You can reproduce any design or drawing using the best pen to plot on your material of choice.

Mechanical Engraving Tool

The next tool that I mounted in the penholder was a mechanical engraving tool. This tool is inexpensive and available at most hardware stores. It is pictured in **Figure 12.7**.

Figure 12.7

Engraving tool.

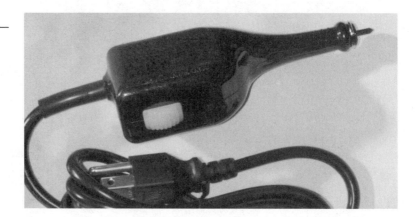

I decided to engrave on a piece of aluminum to determine the smallest size of text the engraver could reproduce with readable text. **Figure 12.8** shows the text sizes engraved in aluminum with

Figure 12.8

Text engraved in aluminum.

a Canadian dime as a size reference. (An American dime is about the same size as the Canadian dime used in the image.)

With the engraver I used, the smallest size of text that can still be read is 6 point. A more accurate engraving tool would most likely be able to produce smaller text. The next image, shown in **Figure** 12.9, is of the same bulldog I plotted using a pen earlier in this chapter, shown engraved on aluminum beside some text.

Figure 12.9

Bulldog engraved in aluminum.

Even though the engraving tool isn't the most accurate or precise available, the CNC machine was able to guide it through the engraving process with a remarkable degree of precision. **Figure** 12.10 shows a closer image of the bulldog and the level of detail achieved by a relatively imprecise tool.

Another vector image I chose to engrave contained a lot of small lines to test the ability of the CNC machine engraver combination. The image in **Figure** 12.11 is of a church engraved in aluminum with a Canadian dime for reference.

Figure 12.10

Close up of engraved
bulldog.

Figure 12.11

Engraved church.

The engraver was also able to etch a design in natural stone. For this experiment, I generated a file to test the engraver with a piece of black granite floor tile. **Figure 12.12** shows the result.

Figure 12.12

Engraving in black granite.

The engraver worked fairly well with the granite, although the lines are a little sloppier than those made in aluminum. All things considered, the machine performed very well as a CNC engraver.

Dremel Tool

The Dremel tool is handy for doing small jobs that require tiny bits. You wouldn't use the Dremel for anything large, but it is great for intricate and small carving. Although it spins quickly, it couldn't take the punishment that a router or larger rotary tool could. There are many bits that can fit its 1/8-inch collet or chuck. I purchased a set of diamond-impregnated bits to try out the Dremel as a glass engraver. Since the bits were inexpensive, I didn't expect a long working life from them. I mounted the Dremel tool in my makeshift holder and set about engraving the Chevy logo in a piece of mirror. In **Figure 12.13** you can see the diamond bit making contact with the surface of the mirror. Thinking that it would

Figure 12.13

Diamond bit engraving glass.

help lubricate the bit, I put a little water on the glass. Although I don't know if the water helped, if I were going to require a diamond bit to do copious amounts of engraving I would build a tub with a pump to provide a constant stream of water to extend the life of the bit.

The diamond bit worked well but didn't last very long; as the tip wore away, the lines it engraved became wider until it stopped cutting into the glass altogether. **Figure 12.14** shows the Chevy logo and the first few engraved letters.

Figure 12.14

Engraving Chevy logo.

Next I used the Dremel to carve the dolphin in pink insulating foam, the kind you can get at the local building supply store. I opened the dolphin image in Profiler6, after I had colored it with shades of gray in CorelDraw and exported it as a jpeg image file. Next, I generated a G-code file that would carve the dolphin. Using an image of the dolphin with a white background so that only the dolphin would be carved, I made the carving seen in **Figure 12.15**. I used a 1/8-inch cutting bit and only overlapped the cuts the minimum that can be set with Profiler6.

Figure 12.15

Carving dolphin in foam.

The dolphin is only 2 inches at its widest point and 5 inches in length. You will notice that the carving is 1/4 inch at the deepest cut, and that the figure is highest at the head, giving the illusion of emerging from the foam. I changed the image of the dolphin, making its body similar shades of gray and the background black. Again I opened the file with Profiler6 and generated G-code to open with KCam. This time I used pine to carve the dolphin. The result of the changes can be seen in **Figure 12.16**.

Figure 12.16

Dolphin carved in pine.

If you generate a G-code file with Profiler6 using more passes and a smaller bit, the resulting carving would look better and wouldn't require as much sanding and hand touchup work.

Master Craft Rotary Tool

This tool is the Canadian Tire version of a rotary cutting tool that is between a Dremel and a router in size and strength. I decided to use this tool to cut hardboard because I didn't want to overtax the Dremel and my router doesn't have a 1/8-inch collet to hold the cutting bit I wanted to use. I drew a crude looking airplane with CorelDraw in three parts, with lines where I wanted the body cut for the wings. I placed the lines for the openings in the body last so they would be cut first. In **Figure 12.17** you can see the resulting airplane and the foam backing I used while cutting.

Figure 12.17

Airplane cut from hardboard.

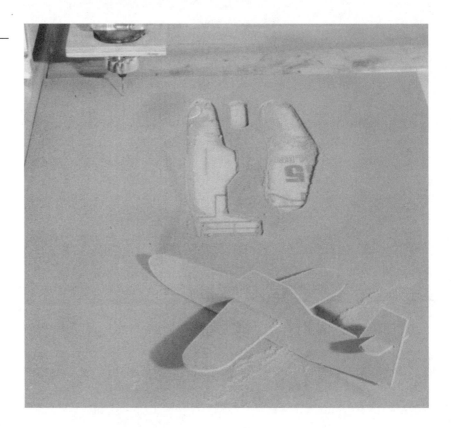

I used the foam to keep the hardboard above the table surface because foam offers very little resistance to the cutting tool. It is also firm enough to maintain the position of the material being cut.

I also acquired a tile-cutting bit that is 1/8-inch in diameter. Normally this type of bit would be used to cut soft-body tile like wall tile. Wall tile has a softer body than floor tile, and wall tile with a white clay body is the softest and easiest to cut. I had some 4-1/4 × 4-1/4-inch wall tile lying around that I used for this example. First I built a little jig to hold the tile in position, as shown in **Figure 12.18**.

Figure 12.18

Jig to hold tile.

The jig is recessed a little less than the thickness of the tile on the three mitered pieces of MDF. The piece of pine at the front of the jig keeps the tile from moving out of the jig and the two pieces screwed to the mitered MDF clamp the tile in place, keeping it from moving up and down. In CorelDraw I made a drawing of a star and a house, both on paper size to match the tile, which I exported as HPGL files. I imported them into KCam and cut the shapes in the tile. In **Figure 12.19** you see the star cut from a white tile.

Figure 12.19

Star cut in tile.

Then I reloaded the jig with a blue tile and cut a star from it, as seen in **Figure 12.20**.

Figure 12.20

Star cut from blue tile.

I did the same with two different colored tiles using the house file, as seen in **Figure 12.21**.

Figure 12.21

House cut from tile.

When you install these tiles, swap the resulting shapes with the opposite colored tile as inserts (see **Figure 12.22**).

Figure 12.22

Swap the shapes of colored tiles.

Router

The router is the most powerful of the tools I will use with the CNC machine and is well suited for larger projects that require the removal of a lot of material. One of the most obvious uses for the CNC machine with a router is sign making. The simplest sign consists of words and numbers and is routed along the path to which the font plots, like the sign in **Figure 12.23**.

Figure 12.23

Simple sign.

This example was routed 1/8 inch deep and the overall size of the sign is 12 × 50 inches. The tallest letter is the G, standing 9-1/2 inches, as depicted in **Figure 12.24**.

Figure 12.24

Tallest letter.

I painted the routed portion orange and didn't worry about getting the paint on the top of the board. After the paint dried, I sanded the entire surface, removing unwanted paint and making the wood ready for a finish. This is the quickest way to make a sign, but if you want to get creative, the next sign I made took a bit more thought to produce. **Figure 12.25** shows a sign with raised letters on a lowered portion of the board with letters routed along single lines at the top of the board.

To create the path for the top letters, I first wrote the text on a page in CorelDraw to correspond to the size of the sign I wanted to make. Then I selected the text and converted it into a bitmap. Once the text is converted to a bitmap you can trace it using CorelTrace. You will need to convert the text bitmap into a black and white image after CorelTrace opens. Trace the image using the centerline method and when you are happy with the result, close CorelTrace. The trace result will now be the selected object on the drawing. Move the trace out of the way so you can select and delete the bitmap you created from the text. Move the trace where you want it. The raised letters on the lowered portion of the sign can be made by placing a box the size of the area to be lowered on the drawing and filling the box with color. Type text over the box, giving the text a different color, and size the text. Because I wanted to use a different router bit for the text at the top of the sign than for the raised letters, I made two files. After I lined everything up I selected the box and letters on the lower portion of the sign and cut them. I exported the remaining part of the sign as an HPGL file named *sign1.plt*. Then I pasted the lower part I had cut from the sign back into the drawing. I selected the top text and cut it from the sign, exporting the balance as *sign2.plt*. When the export dialog opens, go to the last tab and click on simulated fill. I set the width of the simulated fill at 1/4-inch spacing because I intended to use a 1/2-inch router bit to clear out the unwanted material. Set the simulated fill to plot only horizontally. Save the file; now there are two files to use to make the sign. You will need to open the second file with CorelDraw and select all objects, then ungroup them. Remove all the lines within the boundaries of the letters by selecting each one and deleting it. By removing the simulated fill lines from the letters, only the balance of the area will be lowered, leaving the letters raised. When you get your material secured to the table of the CNC machine, just import the first file with KCam and run the program with the first router bit. Then import the second file and change bits before running the program. After I finished the routing, I painted the letters in the top portion of the sign in the routed areas, and in the bottom portion of the sign I painted the lowered areas. When I sanded the sign, I removed paint from the areas around the top letters and from the tops of the bottom letters. See **Figure 12.26** for a closer look.

Figure 12.26

Second sign closeup.

You should be able to make all sorts of different signs once you get CorelDraw figured out. The next example is a prototype of the doors that my friend Geoff wants to make for his kitchen and the original reason why I started thinking about building a CNC machine (see **Figure 12.27**).

Figure 12.27

Prototype kitchen cabinet door.

We used three files to make this door out of MDF cut to size. The first file cleared out the lowered portion of the door by 1/8 inch with overlapping passes of a 1/2-inch bit. Using a box core bit set to cut 1/8 inch deep, the second file routed the lines inside the lowered portion to mimic tongue and groove. A third file was used to run the box core bit around the perimeter of the lowered portion and off the board, while set to cut 1/8 inch from the surface of the door to create the illusion of rail and stile construction. Once we figured out how to make the prototype, the rest of the doors would be easy to make. See **Figure** 12.28 for a close-up of the prototype.

Figure 12.28

Closeup of cabinet door.

I built a small box with one open side as a jig for the next examples (see **Figure** 12.29).

Figure 12.29

Joint jig.

The CNC machine is also well suited for making dovetail and box joints. I have made box joints on my table saw with a jig that took me half a day to perfect, and haven't bothered to make any more jigs, considering the grief the first one gave me. I don't own a dovetail jig for my router, so it seemed like a good time to learn how to make a dovetail joint. I made the box joint first. I had some 1/2-inch pine 5 inches wide, so I created a file in CorelDraw that had lines 2 inches long and spaced 1 inch apart starting from the top of the page, which was the width of my wood, down to the bottom. The 1-inch center will give you 1/2-inch fingers using a 1/2-inch router bit. Make sure you have the page set to landscape when you export any of your files; if you don't, they won't plot correctly. After I screwed my jig to the work surface, I cut a couple of pieces of pine 5 inches long to clamp to the front of the jig, as shown in **Figure 12.30**.

Figure 12.30

Wood clamped to jig.

I drew a line on one of the pieces of wood, 1 inch from the top. This would let me align the wood to the top of the jig, which in theory is parallel to the surface of the work area. I set the zero point of the z-axis at the surface of the ends of the wood clamped to the jig. Next, I set the depth of cut to −.55 inches and travel at .25 inches. See the cut joint in **Figure 12.31**.

Figure 12.31

Box joint cut in jig.

The resulting joint was cut exactly as I wanted, with each finger a little longer than the thickness of the wood, and they fit together perfectly (see **Figure 12.32**).

Next I set up to create a dovetail joint. The dovetail bit that I have is 1/2-inch at the end, tapering down to 1/4-inch. To be able to cut two pieces of wood at the same time, I would need to move the bit through the top of the piece held to the jig vertically, and into the piece held horizontally the thickness of the wood. See the wood clamped to the jig in **Figure 12.33**.

In total, the bit needs to travel 1 inch from the face of the vertical piece and back along the same path before moving to the next cut. To make a joint that will work with this method, the pins and voids need to be the same size. Making them the same size requires that the bit travel on paths that are spaced 3/4 inch apart, as the narrowest part of the bit is 1/4 inch in diameter. To create the plotter

Figure 12.32

Box joint together.

Figure 12.33

Clamped wood for dovetail joint.

file I drew lines that extended into the drawing 1-1/2 inches from the edge of the page and back to the edge, making sure that each line was continuous from start to finish. If the line isn't continuous the bit will raise to the travel height before continuing the line, thus destroying the joint. I made them extend 1-1/2 inches to allow me to set the x-axis zero point to the leading edge of the router bit, being 1/2-inch away from the front of the vertical piece of wood. The 1/2 inch is just for clearance as the bit moves to the next cut. **Figure** 12.34 shows the resulting joint cut in this manner.

Figure 12.34

Dovetail joint cut with CNC machine.

In the future, I'll build a platform that can be mounted on the center frame cross member with the MDF sheets removed from the frame. The limit to the length of a vertically held board will be dictated by how high I build the legs for this machine when I want to get it off the floor.

My last example is how to use this machine to mill wood on a lathe. Since I don't own a lathe that I can put under the table of my machine, I built a makeshift lathe, using MDF, that is powered by my hand drill. The tailstock can be moved and screwed in place to accommodate the material being turned. The live center on the tailstock is made from a bearing and a piece of 1/2 inch ready-rod bolted to a square of MDF, with holes for screws to attach it to the wood to be turned. The driven end of the material is screwed to a piece of MDF that is bolted to a mandrel with a 1/4-inch shaft that fits the chuck of the drill. As always, I hold the drill in place with a large pipe clamp. See **Figure 12.35** depicting the MDF drill-powered mini lathe, with round oak stock attached. I attempted to screw the lathe in place perpendicular to the y-axis (and by default, parallel to the x-axis). The router will only travel down the x-axis, so I set the y position to what I thought was the center of the wood by manually jogging. When I was satisfied with the y-axis position, I zeroed it.

Figure 12.35

Drill powered lathe.

I used oak for this demonstration because I happen to have a bunch of very old logging pikes with 20-foot handles that I haven't found a use for until now. To generate a file with a profile to mill the spinning wood is fairly easy. I don't have any software that will let me draw shapes that only move up and down on the z-axis while traveling only along the x-axis, however. To do this, you would need a CAD program like AutoCAD. I only have CorelDraw, but it turns out that CorelDraw is all I need. To make the profile file, start by creating a long narrow page set to landscape in

CorelDraw. I made my page 1/2 inch high × 15 inches long and then proceeded to draw a profile within the boundaries of the page, as seen in **Figure 12.36**.

Figure 12.36

Drawing a profile in CorelDraw.

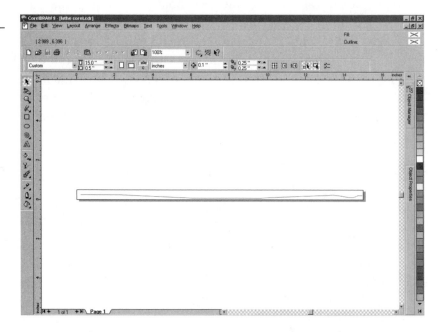

Export this file as an HPGL file. Import the file into KCam. KCam will generate a G-code file that looks like the following:

N000 [KCam Conversion]
N001 [Original File: lathecorel3.plt]
N002 %
N003 G90
N004 M03
N005 G00 Z0.5
N006 G00 X000.000 Y000.000
N007 G01 X000.220 Y000.254
N008 G00 Z-0.25
N009 G01 X002.283 Y000.254
N010 G01 X006.184 Y000.086
N011 G01 X009.745 Y000.086
N012 G01 X012.261 Y000.191
N013 G01 X013.310 Y000.228

N014 G01 X013.413 Y000.233
N015 G01 X013.517 Y000.240
N016 G01 X013.621 Y000.243
N017 G01 X013.673 Y000.242
N018 G01 X013.724 Y000.238
N019 G01 X013.918 Y000.215
N020 G01 X014.005 Y000.202
N021 G01 X014.063 Y000.194
N022 G01 X014.441 Y000.125
N023 G01 X014.727 Y000.125
N024 G01 X014.964 Y000.222
N025 G01 X015.191 Y000.222
N026 G00 Z0.5
N027 G00 X000.000 Y000.000
N028 M05
N029 M30

Notice how the profile is plotted in the plot window of KCam.

You will want to change all the y-coordinates to z-coordinates and this is accomplished using the search and replace function of the G-code editor. After replacing every y with a z, the G-code looks like this:

N000 [KCam Conversion]
N001 [Original File: lathecorel3.plt]
N002 %
N003 G90
N004 M03
N005 G00 Z0.5
N006 G00 X000.000 Z000.000
N007 G01 X000.220 Z000.254
N008 G00 Z-0.25
N009 G01 X002.283 Z000.254
N010 G01 X006.184 Z000.086
N011 G01 X009.745 Z000.086
N012 G01 X012.261 Z000.191
N013 G01 X013.310 Z000.228
N014 G01 X013.413 Z000.233

```
N015 G01 X013.517 Z000.240
N016 G01 X013.621 Z000.243
N017 G01 X013.673 Z000.242
N018 G01 X013.724 Z000.238
N019 G01 X013.918 Z000.215
N020 G01 X014.005 Z000.202
N021 G01 X014.063 Z000.194
N022 G01 X014.441 Z000.125
N023 G01 X014.727 Z000.125
N024 G01 X014.964 Z000.222
N025 G01 X015.191 Z000.222
N026 G00 Z0.5
N027 G00 X000.000 Z000.000
N028 M05
N029 M30
```

You need to do a little more editing. In line N006 change the Z to a Y. Remove line N008 and change the Z in line N027 to a Y. You only want movement on the z- and x-axis; the y-axis must remain stationary; that is why the y-axis was zeroed at the center of the piece to be milled and any reference to the y-axis in the G-code must be only 000.000. Remove the line numbers with the editor and then add the numbers back again. After final editing it will look like this:

```
N001 [KCAM CONVERSION]
N002 [ORIGINAL FILE: LATHECOREL3.PLT]
N003 %
N004 G90
N005 M03
N006 G00 Z0.5
N007 G00 X000.000 Y000.000
N008 G01 X000.220 Z000.254
N009 G01 X002.283 Z000.254
N010 G01 X006.184 Z000.086
N011 G01 X009.745 Z000.086
N012 G01 X012.261 Z000.191
N013 G01 X013.310 Z000.228
N014 G01 X013.413 Z000.233
```

N015 G01 X013.517 Z000.240
N016 G01 X013.621 Z000.243
N017 G01 X013.673 Z000.242
N018 G01 X013.724 Z000.238
N019 G01 X013.918 Z000.215
N020 G01 X014.005 Z000.202
N021 G01 X014.063 Z000.194
N022 G01 X014.441 Z000.125
N023 G01 X014.727 Z000.125
N024 G01 X014.964 Z000.222
N025 G01 X015.191 Z000.222
N026 G00 Z0.5
N027 G00 X000.000 Y000.000
N028 M05
N029 M30

Recompile the G-code, and from the top the profile will be a straight line along the x-axis. Look at the plot from the front and it's now what you need to mill the profile with a lathe. If you want to set the top of the material to be milled as zero, offset the z-axis until the profile is just below the surface of z-axis zero, as seen in **Figure 1**2.37.

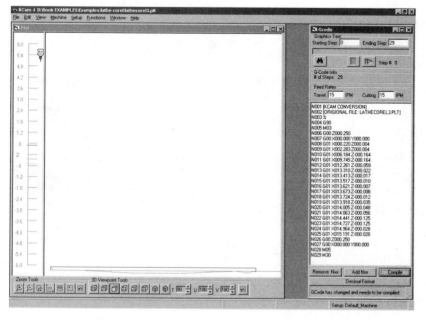

Figure 12.37

Offsetting the z-axis.

I moved the profile down about 1/16 inch at a time because I didn't want to stall my homemade lathe. I used a V groove router bit to mill the wood because it was sharp and for no other reason. You can see the CNC machine milling wood on the lathe in **Figure 12.38**.

Figure 12.38

Lathe turning wood under router.

The finished product is shown in **Figure 12.39**.

Figure 12.39

Finished turning.

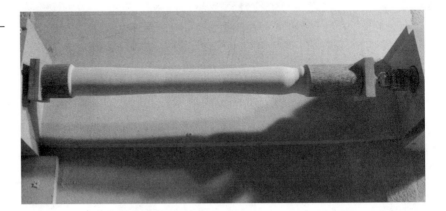

I could probably write an entire book on how to use this machine but a chapter will have to suffice. I suspect that those who build a Workshop Bot for themselves will find hundreds of uses, and enjoy inventing new ways to use their machines every day.

Sources
of Material

Electronic Components

L297/L298 Integrated Circuits

Manufactured by STMicroelectronics

http://us.st.com/

The following list of distributors comes directly from the STMicroelectronics Web site. They have distributors in almost every country on this planet so I've only listed a few for Canada and the United States. Visit their site to find the distributor nearest you.

CANADA

Arrow
http://www.arrow.com
Phone: 613-271-8200
Fax: 613-271-8203
Phone: 905-670-7769
Fax: 905-670-7781

AVNET

http://avnet.com

Phone: 613-226-1700

Fax: 613-226-1184

Phone: 905-812-4400

Fax: 905-812-4458

Future Electronics

http://www.futureelectronics.com

Phone: 905-612-9200

Fax: 905-612-9185

Phone: 613-727-1800

Fax: 613-727-9819

Pioneer Standard

http://www.pioneerstandard.com

Phone: 613-226-8840

Fax: 613-226-6352

Phone: 905-405-8300

Fax: 905-405-6423

Phone: 519-672-4666

Richardson Electronics

http://www.rell.com

Phone: 514-939-9640

UNITED STATES, CALIFORNIA

Arrow

Internet: www.arrow.com

Phone: 619-565-4800

Fax: 619-565-2959

Phone: 818-880-9686

Fax: 818-880-4687

Phone: 949-587-0404
Fax: 949-454-4206

AVNET
Internet: avnet.com
Phone: 818-594-0404
Fax: 858-385-7500
Phone: 949-789-4100

Future Electronics
Internet: www.futureelectronics.com
Phone: 949-453-1515
Fax: 949-453-1226
Phone: 619-625-2800
Fax: 619-625-2810

Mouser Electronics
Internet: www.mouser.com/stmicro
Phone: 800-346-6873
Fax: 619-449-6041

NU Horizons Electronics
http://www.nuhorizons.com
Phone: 949-470-1011
Fax: 949-470-1104
Phone: 619-576-0088
Fax: 619-576-0990
Phone: 805-370-1515
Fax: 805-370-1525

Pioneer Standard
http://www.pioneerstandard.com
Phone: 949-753-5090
Fax: 949-753-5074

Phone: 619-514-7700
Fax: 619-514-7799
Phone: 818-865-5800
Fax: 818-865-5814

Richardson Electronics
http://www.rell.com
Phone: 818-594-5600
Fax: 818-594-5650
Phone: 909-600-0030
Fax: 909-600-0064

Lineal Motion

The companies listed manufacture components used to achieve lineal motion like bearings and guide rails or acme and ball screws. Although I used INA bearings and guide rails for my machine, you can use whichever product you want. This list is short and doesn't include every manufacturer of these products. You will find that most of the companies on this list are represented all over the world. Canadian Bearings is my local source for lineal motion components; they carry many of the parts produced by the manufacturers on the list.

Distributor

Canadian Bearings Ltd.
500 Trillium Drive
Kitchener, Ontario
Canada N2R 1A7
Tel: 519-748-5500
Fax: 519-748-5040
Toll Free: 1-800-265-8206
After Hours: 519-575-2705
http://www.canadianbearings.com/

Manufacturers

Bishop-Wisecarver Corporation
2104 Martin Way
Pittsburg, CA 94565-5027
Telephone: 925-439-8272
Toll Free: 888-580-8272
FAX: 925-439-5931
http://www.bwc.com/html/index.html

INA Canada, Inc.
2871 Plymouth Drive
Oakville, Ontario
Canada L6H 5S5
Tel.: 905-829-2750
Fax: 905-829-2563
http://www.ina.com/

INA USA Corporation
308 Springhill Farm Road
Fort Mill, SC 29715
Tel.: 803-548-8500
Fax: 803-548- 8599
http://www.ina.com/

INA Linear Technik
A Division of INA USA CORPORATION
3650 D Centre Drive
Fort Mill, SC 29715
Tel.: 803-802-0511
Fax: 803-802-0636
http://www.ina.com/

NSK Americas
Head Office 4200 Goss Road
Ann Arbor, MI 48105
Tel: 734-913-7500
http://www.nsk.com/

NSK Canada
Toronto/Head Office
5585 McAdam Road
Mississauga, Ontario
Canada L4Z 1N4
Tel: 905-890-0740
Fax: 905-890-0434

SKF
AB SKF, SE-415 50 Göteborg, Sweden
Hornsgatan 1
Tel: +46-31-337-10-00
Fax: +46-31-337-28-32
http://www.skf.com

SKF Canada Limited
40 Executive Court
Scarborough, Ontario
Canada M1S 4N4
Tel: 416-299-1220
Fax: 416-292-0399
http://www.skf.ca

SKF USA Inc.
1111 Adams Avenue
Norristown, PA 19403-2403
Tel: 610-630-2800
Fax: 610-630-2801
http://www.skfusa.com

SKF Motion Technologies
1530 Valley Center Parkway Suite 180
Bethlehem, PA 18017
Tel: 800-541-3624
Fax: 610-861-4811
http://www.linearmotion.skf.com/

THK Canada
130 Matheson East, Unit 1
Mississauga, Ontario
Canada L4Z 1Y6
Tel: 905-712-2922
Fax: 905-712-2925
http://www.thk.com/

THK America, Inc.
Head Office
200 E. Commerce Drive
Schaumburg, IL 60173
Tel: 847-310-1111
Fax: 847-310-1182
http://www.thk.com/

Thomson Industries, Inc.
Corporate Headquarters
2 Channel Drive
Port Washington, NY 11050
http://www.thomsonind.com/default.htm

USA, Canada, or Mexico
Phone 1-800-554-8466
Fax 1-516-883-9039

Europe
Phone 44-1271-334-500
Fax 44-1271-334-502

UK
Phone 0800-975-1000
Fax 0800-975-1001

France
Phone 0800-90-5721
Fax 0800-91-6315

Germany
Phone 0800-1- 816-553
Fax 0800-1-816-552

Elsewhere
Phone 1-516-883-8000
Fax 1-516-883-7109

Stepper Motors

Princess Auto
P.O. Box 1005
Winnipeg, Manitoba
Canada R3C 2W7
Ph: 204-667-4630
Fax: 204-663-7663
In Canada Call Toll Free:
1-800-665-8685
Fax: 1-800-265-4212
http://www.princessauto.com/

Pacific Scientific
4301 Kishwaukee Street
PO Box 106
Rockford, IL 61105-0106
Phone 815-226-3100
Fax 815-226-3148
http://www.pacsci.com/

Sanyo Denki America, Inc.

468 Amapola Avenue,

Torrance, CA 90501

Phone: 310-783-5400

Fax: 310-212-6545

http://www.sanyo-denki.com/

Metal

If you need a small quantity of metal and can't find it at a scrap yard, then a good place to look is a store called Metal Supermarkets. They have shops in Canada, the United States, England, Scotland, and Austria. Go to their Web site at http://www.metalsupermarkets.com/ to find a local store.

Metal Supermarkets (Kitchener/Waterloo)

5 Forwell Road, Unit 4

Kitchener, Ontario

Canada N2B 1W3

Tel: 519-742-8411

Fax: 519-742-9377

Toll Free: 800-742-8620

Index

About the Author

Geoff Williams is a woodworking enthusiast who owns and operates an athletic flooring company in Ontario, Canada. Also a professional photographer, he has extensive experience troubleshooting and repairing printed circuit boards.